FROM PASCAL TO FORTRAN 77
APPLICATIONS FOR SCIENTISTS AND ENGINEERS

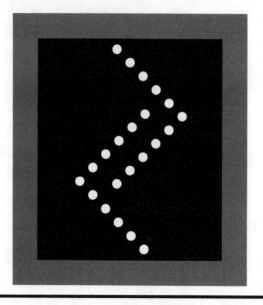

DAVID DONALD MILLER

Bemidji State University

HARCOURT BRACE JOVANOVICH, PUBLISHERS
and its subsidiary, Academic Press
San Diego • New York • Chicago • Austin
London • Sydney • Tokyo • Toronto

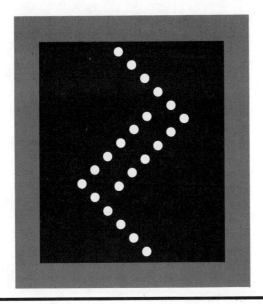

FROM PASCAL TO FORTRAN 77
APPLICATIONS FOR SCIENTISTS AND ENGINEERS

PREFACE

In recent years, the popular computer science programming textbooks have been gradually swinging toward Pascal, primarily because computer science concepts are easily taught and learned using Pascal. But other disciplines, in particular engineering and the physical sciences, require a knowledge of programming. In the engineering work-a-day world, programming is firmly entrenched in FORTRAN for the following reasons:

- Because of its simple structures, FORTRAN is easy to use in some applications.
- FORTRAN offers mathematical sophistication not available in standard Pascal or in any Pascal extensions.
- FORTRAN supports a wide variety of file manipulation constructs not available in standard Pascal.

- A real-world application often involves a combination of several subprograms, which is supported by the FORTRAN system but not by Pascal.
- Because of its longevity, millions of lines of FORTRAN already exist and must be maintained (modified, added to, and used).
- FORTRAN compilers are usually error free and generate optimized code.

From Pascal to FORTRAN 77: Applications for Scientists and Engineers is directed to those students who have been introduced to programming through Pascal but whose career goals are in engineering or the physical sciences. Such students will need a working knowledge of FORTRAN. This textbook is also directed to practitioners who must make a transition from Pascal to FORTRAN.

FORTRAN is a programming language of the ancients—it could even be called a classical language, since in the world of computers 30 years is quite a long time. And, just like an old house, it has acquired a whole attic full of elaborate rules and special cases. Many of the original FORTRAN constructions have been replaced over the years, because we know more about compiler languages now than we did when FORTRAN was pioneering the field. This book cuts through the historical bric-a-brac and concentrates primarily on the most useful parts of FORTRAN. Incidentally, many of the best parts of FORTRAN have Pascal counterparts as well, and I will draw heavily on the student's Pascal knowledge to show, by example and counter-example, FORTRAN and Pascal code side by side. The older FORTRAN constructions have been included, for the sake of completeness, in an appendix, because students will surely see older versions of FORTRAN in their careers.

HOW TO USE THIS TEXTBOOK

The material moves along quickly, especially through the introductory chapters. In fact, the student will be coding simple programs at the end of the first chapter and writing serious FORTRAN by the end of the third chapter. Each chapter contains the following features:

- A discussion of the chapter material, often including FORTRAN program segments and the equivalent Pascal segment for comparison.

- An example, including explanation, algorithm, commented code, sample execution, and testing suggestions.
- "Pitfalls" that a Pascal programmer and a novice FORTRAN programmer would be likely to encounter.
- Exercises that elaborate on the example and that introduce related topics.

The text makes the following assumptions about the student's level of computer expertise:

- It assumes the student has already been introduced to such primary programming topics as basic computer architecture, interactive terminal sessions, program structure, and the like. The student's time will not be wasted by the introduction of that kind of information again, and the text will, instead, draw upon that knowledge and extend it.
- It assumes the student is acquainted with problems that occur in the physical world—some of the examples may have been seen in other courses.
- Since this is a textbook for engineers and other physical scientists, most of the examples and exercises are drawn from the scientific and mathematical world; the student will never have to compute a payroll. A few examples involve games, since programming should be fun too. Some problems involve mathematics through calculus and, in some cases, differential equations. But higher math is not normally required, especially early in the text. The intention is to introduce real-world problems, approaches, and software solutions, and in doing so, to introduce topics like modeling functions, solving systems of equations, and numerical integration. All of the students' acquired mathematical skills should be brought into focus in this course.

The last point requires some elaboration: Computer generated answers tend to be believed without supporting evidence. This is an extremely dangerous attitude and one that should be corrected. A secondary purpose of the text is to illustrate some introductory applications of digital computing in a practical world. This is not supposed to be a numerical analysis course, but it may well be the only exposure some students will ever have to the subject. I am concerned that students learn both the possibilities and the impossibilities of digital computation, and that they be made aware of the misapplication of computers as well as their applications. I have tried to impart a "feeling" for detecting unreasonable answers—and have listed some heuristics that may help students examine their code when they are suspicious of their answers.

One other thing: This book will force students to go back into some of their "old" textbooks. Maybe they thought they would never have to open them again, but I want them to know that education in a technological world is never ending: Those texts are still valuable. Several of the exercises are designed to send students into those other books to look up derivations and formulas: I have not simply provided the necessary equation as part of the exercise. That is how it is done in industry: Employees are given a problem statement and it is up to them to define an approach and do the necessary research.

This book concentrates on the standard, FORTRAN 77. Many of the examples here are coded using VAX FORTRAN in the "standard" FORTRAN 77 mode. Several of the examples are also coded using MS-DOS FORTRAN on the HP 150 Touch Screen. MS-FORTRAN is based on the "subset" standard FORTRAN 77, which in this case also includes many "full" standard features. It has its shortcomings, however, and when it is appropriate, I warn the student of any implementation-specific snags that I have discovered in the process. Maybe they will not have any trouble on their particular machine and maybe they will have other problems.

ACKNOWLEDGMENTS

I wish to thank the following reviewers for their helpful suggestions: Beverly Bilshausen, College of DuPage; Mark J. Christiansen, Georgia Institute of Technology; Larry Cottrell, University of Central Florida; Charlotte Fischer, Vanderbilt University; Charles Gould, Florida Institute of Technology; Lawrence O. Hall, Florida State University; Larry Neal, Eastern Tennessee State University; Charles Neblock, Western Illinois University; Bob Pabasco, University of Idaho; Jim Pannell, DeVry Institute; Paul Paulson, Central Michigan University; Don Ramsey, Tennessee Technological University; Paul Ross, University of Millersville; Richard Sleight, University of Washington; and Terry Smith, Northeastern University.

DAVID DONALD MILLER

CONTENTS

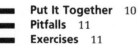

FROM PASCAL TO FORTRAN 77
APPLICATIONS FOR SCIENTISTS
AND ENGINEERS

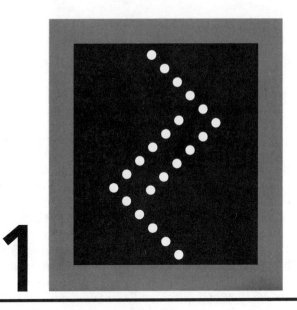

INTRODUCTION TO FORTRAN 77

FORTRAN is a formal computer language introduced in the late 1950s. The computer science field was in its infancy then. You wouldn't even recognize the vocabulary: A "computer" was the term used for what we call a "programmer" today. An "electronic calculator" was the word used for what we now call the "computer." The primary function of the "electronic calculator" was numeric applications. But the industry that had begun in the 1940s was growing; there was an increasing need for "automatic programming," the term first used for "compilers." In 1954, John Backus and three associates produced a preliminary report for the "IBM Mathematical FORmula TRANslator System," FORTRAN.

1.1 A BRIEF HISTORY OF FORTRAN

By the late 1950s computer hardware was becoming more complex. IBM's "new" 704 was going to introduce "built-in" real arithmetic and indexing. Previously, both these functions were performed in software. Since hardware was advancing in complexity, it was only natural to expect that software would be advancing too. In fact, a software revolution was in the making and, of course, there would be mind sets to overcome. Because FORTRAN was intended as a replacement for the IBM 704 assembler, it was designed to be almost an instruction-by-instruction substitution system, except in the areas of expression analysis and indexing. Backus didn't expect to set the industry on fire; rather, he intended to make it easier to program the 704. He fully expected other manufacturers to develop similar systems—just as they had developed assemblers specifically for each machine.

This first version of FORTRAN was distributed to IBM installations in early 1957, but there were shortcomings (and bugs, too, no doubt) with that initial version. So, in the fall of 1957, FORTRAN II was begun. It was distributed in the spring of 1958 and, by the end of that year, sixty IBM installations were using it. From that time, FORTRAN became an institution, not an experiment, within IBM. Its success bred new ideas and new bureaucracies. FORTRAN II spawned ideas for FORTRAN III (never released), which in turn became a stepping stone to FORTRAN IV, designed in 1962. Much to Backus' surprise, in the early 1960s the rest of the computer industry got on the FORTRAN bandwagon. The American National Standards Institute (ANSI) began to define a "standard" FORTRAN in 1962. This work was completed in 1966 and, predictably, that version was called FORTRAN 66. During this standardization process, many dialects of FORTRAN grew, partly to strengthen the language and partly to make it less IBM-dependent. So, a second attempt to standardize was completed in 1977. Of course, it was called FORTRAN 77. There is yet another standardizing attempt taking place right now—expected to become FORTRAN 88.

Other "automatic programming" systems were developed during those early days of FORTRAN, notably ALGOL, the forerunner to Pascal. But IBM dominated the computer hardware industry through the mid-1960s and FORTRAN belonged to IBM; therefore the majority of the computer industry used FORTRAN. It is interesting to note that IBM introduced PL/1 into a more competitive industry in 1966, but that language never came close to having the popularity of FORTRAN.

Why has FORTRAN endured? Because of nostalgia or tradition? Perhaps in part, but that isn't the whole story. The original reasons for FORTRAN are still valid: It is a simple language that is very powerful for mathematical applications. It has more capabilities in that area than Pascal because of double precision and complex data types, its extensive library of intrinsic functions, and its very powerful I/O features. And that is in fact our focus: to help you get to know FORTRAN as a tool for solving problems that are modeled by mathematics.

1.2 LINE FORMAT

The first thing we are going to talk about requires that you realign your thinking. FORTRAN is not a format-free language like Pascal is. A FORTRAN program contains "statement" lines and "comment" lines. The line concept is really an outgrowth of the IBM card, and much of what we're going to say stems from that mind set. Ever wonder why a CRT terminal is 80 characters wide? It's the width of an "IBM card," of course. This is a good example of how we get locked into historic standards. In those days, programmers carried cards, and it was common practice to number them sequentially (just in case the deck of cards was dropped). Columns 73 through 80 are reserved for a sequence number. A FORTRAN 77 line can have, at most, 72 characters. Any statement longer than that must be continued on the following line, but that second line must have a "continuation" character in column 6. Let's look in detail at these format rules.

Rules for comment lines are easiest, so we will deal with them first: If the character in position 1 of the line is either a "*" or a "C", the entire line is considered to be a comment. There is no way to embed comments within a statement, as Pascal allows. Comment lines cannot be continued. A line with nothing on it is treated as a comment line—it makes a good spacer.

Statement lines are divided into three "fields." They are called the "label," "continuation," and "statement" fields. Perhaps a picture will better illustrate these definitions:

```
          1                 2                 3                 4  . . . . .  7                 8
1234567890123456789012345678901234567890012. . . . .8901234567890
```

Label

Continuation

Statement

The "label" field occupies positions one through five. We will discuss labels in Chapter 3. A continuation line cannot contain a label.

The "continuation" field is position 6, as mentioned. Any statement that extends beyond position 72 must be continued on the following line. A continuation line is identified by any character (except zero or blank) in the continuation field. Many programmers use a plus (+) symbol as a continuation flag. In the old days, we would use a number in the continuation field, as a sort of mini-sequence number—again, worrying about dropping the deck. We worried about that a lot! Batch systems were all we had, and the card deck was handled by several people because so many operations were performed manually.

The statement field is positions seven through 72. Anything extending beyond position 72 is ignored and not even flagged as an error. If you are careless and extend your statement beyond 72, you may not even get an error message because if the statement is syntactically correct up to position 72, no error occurs. But, let's say there is a closing parenthesis in position 73; then you get a nasty message from the compiler, but when you list your file, nothing looks wrong because the closing parenthesis is there. So, be very careful with long statements.

Here is an example. I've brought along the "ruler" so you can see which columns I'm working in. This program isn't supposed to make sense (although it will run); it is intended to illustrate statement format only.

```
         1         2         3         4  .....  7         8
123456789012345678901234567890123456789012.....8901234567890

       PROGRAM I01
C
C Sample program showing statement formats
C
       WRITE (*,*)
     1     'Input value for I'
       READ (*,*) I
100    CONTINUE
       WRITE (*,*) 'I is', I

       END
```

Position numbers are a little hard to work with on terminals, so the various manufacturers have developed shortcuts. Students will have to discover how these shortcuts have been implemented on their host system.

Now we'll move on so you can get to coding and discover these problems yourself. You may want to re-read this section later; you will appreciate it more fully.

1.3 PRIMARY "PROGRAM" STRUCTURE

In concept, a FORTRAN program is similar to a Pascal program. The details involved in writing a program, however, are vastly different: FORTRAN is "upside-down"; that is, the main program is entered in the beginning of the file, before any of the subprograms—rather than last, as in Pascal. Another difference we will investigate later is that subprograms cannot be nested and can be in any order in the file—rather than "bottom-up," as Pascal demands.

The first structure we will examine in detail is the main program. It is diagrammed below:

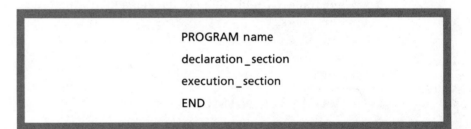

```
PROGRAM name
declaration_section
execution_section
END
```

where the following definitions apply:

- "name" is a variable name as defined in Section 1.4.
- "declaration_section" is introduced in Section 1.5, and discussed more completely in Chapter 4.
- "execution_section" is introduced later in this chapter and in following chapters.

A Pascal programmer will immediately notice the following differences:

a) There is no punctuation (semicolon or period) on the PROGRAM or the END lines.

b) There are no file specifications on the PROGRAM line.

c) There is no keyword (like the Pascal BEGIN) separating the "declaration_section" and the "execution_section."

Many constructs contain optional elements; these are historic. They didn't occur in the early versions of FORTRAN, but were added later. For instance, as the language was developed, it was decided to make things like the PROGRAM construct optional so that software written for the earlier versions of FORTRAN would still compile under later versions. This is often confusing to the FORTRAN novice, and we are going to see this sort of thing again and again, so be prepared. However, we will try to make things easier by extending your Pascal habits. Pascal was created, in part, to make language constructions easier to anticipate (because they are consistent). We will do the same thing with FORTRAN, dwelling only on the consistent aspects of the language—as much as we are able.

As you see, this PROGRAM form is similar, but not identical, to Pascal. So, as you write your first FORTRAN program at the end of this chapter, be forewarned: FORTRAN isn't a trivial subset of Pascal. Rather, Pascal and FORTRAN have subtle differences as well as similarities. As we continue, we will point out the FORTRAN pitfalls common to Pascal programmers.

1.4 VARIABLE NAMES

FORTRAN variable names will look familiar to you. However, unlike Pascal, they can be at most six characters long. As you may have begun to suspect by now, a lot of unusual features in FORTRAN are actually carry-overs from the original FORTRAN implementations. At one time there was a valid reason for this six-character restriction; it has long since disappeared. FORTRAN was invented for a 36-bit machine—the IBM 704 and, in those days, a byte was six bits. FORTRAN I was designed so that a variable name would fit in only one machine word.

As with Pascal variables, a FORTRAN variable name must begin with a letter, and it may contain numbers. It can be, at most, six letters and numbers long, so variable names like TOTAL, CURENT, and AREA2 are all valid. But, COUNTER, VARIABLE, and DIFFERENCE are not valid variable names because of their lengths.

Here is one more historical note. You will find that there are no reserved words in FORTRAN. This means that IF and DO are valid variable names too. I mention it because you may run across this in practice sometime; don't be confused by it. However, using this feature in your programs is to be strongly discouraged.

1.5 VARIABLE TYPES

The concept of data "typing" was unknown in the early days of FORTRAN. Until FORTRAN IV, it was unnecessary (and in fact, impossible) to declare a variable to be, for instance, an integer. Remember that FORTRAN was to be a numerical tool and therefore had to deal primarily with real numbers.

But even the earliest versions of FORTRAN could do both double precision and complex operations. The method for implementing this concept is worth mentioning in passing. The arithmetic expression that was supposed to be evaluated using, for instance, complex arithmetic, was flagged with an "I" in column 1 (the "comment" column). The "I" flag would cause the compiler to generate the proper code for that single statement. Double precision was implemented in the same way using a "D" in column 1. It was the programmer's responsibility to declare an array (using a DIMENSION construct) for the specific variables, which had to be real, in the double precision or complex construct. Chapter 8 deals with the details of using complex arithmetic in FORTRAN 77.

From the earliest version of FORTRAN, all variables had implicit data types and this "feature" has been carried into FORTRAN 77. Implicit typing has many drawbacks, but the chief one is that it means there are only two data types: INTEGER and REAL. Here is how it works: Any variable whose name begins with the letter I, J, K, L, M, or N is considered to be an integer and all others are assumed to be real variables. For instance, LOOPER, MATRIX, and N are assumed to be integer variables. So, many FORTRAN programs end up containing strange variable names because COUNT and START will be real variables, but ICOUNT and KSTART will be integer variables. The unusual spelling is a result of implicit typing. The programmer is forced to use unusual spelling variations to create a variable of the proper type and still retain some meaning. We now call FORTRAN's implicit typing the I-N rule. This idea will be developed further in Chapter 4.

1.6 CONSTANTS

Pascal experience with constants carries over to FORTRAN very nicely. FORTRAN uses the same format for integers and reals.* For instance, 137, −42, and 0 are all integers and 0.1234, 78.9E+20, and −543.1 are reals. Remember, FORTRAN was born in an era that used computers to solve numerical problems, so there are also "double precision" and "complex" data types in FORTRAN. The value of complex arithmetic is probably already apparent to you, and you will soon learn about extended precision. These topics will be covered in greater detail in later chapters.

FORTRAN also supports "Character Strings" which Pascal programmers would call PACKED ARRAY OF CHARs. A nice FORTRAN feature is that, when dealing with strings, you don't have to fool around with ARRAYs. We will see more examples as we go along, but it shouldn't surprise you to see that 'Total Volume' and 'BEMIDJI STATE UNIVERSITY' are legal string constants.

FORTRAN also supports something similar to Pascal's "Boolean" type, called LOGICAL constants. These constants take the form of .TRUE. and .FALSE..

1.7 ARITHMETIC OPERATIONS AND ASSIGNMENT CONSTRUCTS

Arithmetic operations (+, −, *, and /) will be familiar to Pascal programmers, with only a few exceptions listed below. The precedence of these operations is: "**" (exponentiation) highest, "*" (multiplication) and "/" (division) next, and "+" (addition) and "−" (subtraction) lowest.

a) Unlike standard Pascal, there is an operator to raise a number to a power (exponentiation). It is "**", and, although some versions of Pascal do allow this (e.g., DEC®), it is not part of Pascal's formal definition. But, B**2−4*A*C is an acceptable expression in FORTRAN.

b) FORTRAN provides no special integer division operator like Pascal's DIV. In FORTRAN, if either operand is REAL, the result is REAL too. If both operands are INTEGER, the result is an

---

*Originally, the term "floating" or "floating point" was used instead of "real." So, if you ever hear any grey-haired programmers using either term, you will know what they are talking about.

INTEGER. Here is an example: 1/4 is zero in FORTRAN because both constants are integers (whereas it would be 0.25 in Pascal). In FORTRAN if either operand is REAL, then real division is performed. For example, 1.0/4 is 0.25.

c) The Pascal equivalents of MOD and ROUND are available in FORTRAN too, but are implemented differently. We will get around to showing some examples in Chapter 5.

You should have no problems with arithmetic operation precedence—it is exactly the same as Pascal, with the exception that in FORTRAN exponentiation has the highest precedence. But remember: When in doubt, add parentheses for clarity. Always keep in mind that you might have to read your own code someday, and I've found that the more comments and other helps I give myself, the better.

We now have all the pieces to make a simple FORTRAN assignment statement. As with Pascal, the expression is on the right and a variable name is on the left of the assignment operator. There are two differences, however:

a) The FORTRAN assignment operator is " = " (not Pascal's " := ").

b) There is nothing to separate a FORTRAN statement from its neighbor like Pascal's ";". Instead, the "continuation" concept is used. Each FORTRAN statement starts on a new line except when the continuation field is used.

As an example, the FORTRAN statement to compute the one root of a quadratic equation would be written:

```
X1 = ( -B + SQRT ( B**2 - 4*A*C) ) / ( 2*A )
```

Notice I have added blanks here and there to improve readability, but there is no insistence on blanks anywhere in the expression. FORTRAN is very easygoing on blanks: *it doesn't require* them anywhere, even to separate keywords from variables. For instance, ELSEIF and ELSE IF and E L S E I F all mean the same thing in FORTRAN. In fact, I've seen programs written with blanks inserted to confuse the unsuspecting reader—this is, of course, an elaborate jest, and not encouraged. Again, we would like to caution you not to abuse this feature of FORTRAN or else you will create unreadable code.

An integer assignment example would be:

```
KOUNT  =  KOUNT  +  1
```

This has the same functional meaning as its counterpart in Pascal.

1.8 INPUT AND OUTPUT CONSTRUCTS

The easiest-to-use forms of input and output constructs are similar to Pascal's. More complex forms are covered in Chapters 6 and 10. The general forms of these easy constructs are:

> READ (*,*) input_list
>
> WRITE (*,*) {output_list}

where the following definitions apply:

- "input_list" consists of one or more variable names, separated by commas.
- "output_list" has zero or more variable names, expressions, and/or character string constants in it.

Here is the special case of the WRITE statement with an empty list:

```
WRITE (*,*)
```

which works just like the Pascal statement:

```
WRITELN
```

that is, it skips a line when the "output_list" is empty.

Put It Together

Finally, here is a very simple program. It prompts for a value for "I", reads it, then prints it:

```
PROGRAM IO1

WRITE (*,*) 'Input value for I'
READ (*,*) I
WRITE (*,*) 'I is', I

END
```

This PROGRAM will produce the following output:

```
Input value for I
1234567
I is        1234567
```

Pitfalls

This section will be included in every chapter. It is designed to show the student—especially the Pascal programmer—what errors are likely to be encountered when first applying the concepts just introduced.

1. You may find yourself using ":=" instead of "=" in an assignment statement.

2. You must be very careful of the statement position; don't start before position 7 or go beyond position 72.

3. Remember the I-N rule will automatically "type" your variables. You should also be aware that misspellings will not be detected by FORTRAN, but will be automatically declared.

4. Don't use ";", but on the other hand, if your statement extends beyond position 72, you must use the continuation feature.

Exercises

By this time I'm sure you are anxious to try to "beat the compiler," so here are a few simple programs you can use to try to sharpen your newly discovered skills.

1. Design/code a program using *several* WRITE statements; display your name, address, phone number, and student ID number.

2. Using a *single* WRITE statement, repeat the problem above. You should use this problem to try to figure out how a continuation statement works.

3. Can you embed an arithmetic expression, like 3∗I, in the FORTRAN WRITE statement, as you can with Pascal's WRITE? Design/code a trivial example, or counter-example, to answer this question.

4. Investigate internal representation of integers. What is the largest, and the smallest, number recognized by your computer? Design/code a program that will:

   a) READ an integer (as in the previous section),
   b) then WRITE it.

   Observe the results. Is the output number the same as the number you entered? If so, run your program again with a different number. If the two are not the same, you have found a number that is either too big or too small for the computer. Hint: Your computer is a binary machine—you only have to try numbers that are related by powers of two. For instance, start with 1 and double it. Likewise, start with −1 and double it.

5. Investigate the computer representation of real numbers. Initialize a bunch of irrational REAL variables, like PI, E, etc., to their respective values and then print them. Which ones convert properly? Are the lowest order digits correct? What about 10∗PI and 100∗PI?

6. Sum a fraction, like 1/11, 11 times using the following as a guide:

```
FRAC = 1.0/11.0
SUM = FRAC + FRAC + FRAC + ... + FRAC
WRITE (*,*) SUM-1.
```

   Is the sum unity? If it is, try again with a larger denominator. Eventually, the sum will be other than unity. Why?

7. 1/10 is an irrational binary number—is that surprising? What is its binary representation? This problem is to be done without the computer to help you. Hint: You might have to dig out another text to review this concept—do you remember how to change number bases? This is a great opportunity to review that principle.

8. As a continuation of the previous problem, a more challenging
question is this: Design/code a program with the statements

```
A = 1.0/10.0
WRITE (*,*) A
```

in it. Since you know that the conversion is irrational, why is
the value printed rational and correct? What do you suppose is
going on?

2

CONDITIONAL
CONSTRUCTS

Conditional execution of statements is controlled by the IF construct in FORTRAN, just as it is in Pascal. However, FORTRAN has three different forms of the IF but only one of them has an ELSE. The reason for these several flavors is historical. However, over time, the Pascal version of the IF-THEN-ELSE was included. In FORTRAN 77, this form is called the "Block" IF. It is the only variety we shall study. The other versions of the IF are certainly in use today, but they don't lead to good programming practices. We will use only the form introduced in this chapter.

You will notice that, beginning with this chapter, all examples are "stylized" or "prettyprinted." This is not a FORTRAN-imposed rule, but rather, my opinion of what code should look like to be readable. You may have encountered "indented" Pascal too. My rules are simple:

1. All statements under an IF, ELSEIF, or ELSE are indented four positions.
2. All statements in a DO loop are indented four positions.

These rules are applied to nested structures also: DOs within IFs, DOs within DOs, IFs within DOs and IFs within IFs. There is no way this practice can be forced on you—especially if you've never seen Pascal "prettyprinted." We can only suggest that you try it out; see if it makes your program easier to read, understand, and debug.

2.1 CONDITIONAL CONSTRUCTS

The most general form of the Block IF is constructed in the following manner:

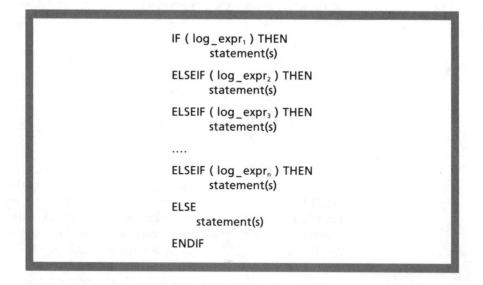

```
IF ( log_expr₁ ) THEN
        statement(s)

ELSEIF ( log_expr₂ ) THEN
        statement(s)

ELSEIF ( log_expr₃ ) THEN
        statement(s)

....

ELSEIF ( log_exprₙ ) THEN
        statement(s)

ELSE
        statement(s)

ENDIF
```

where the following definitions apply:

- "log_expr_i" are logical expressions, as defined in Section 2.2.
- "statement(s)" is one or more complete FORTRAN statements.

Let's examine this general statement by looking at the similarities and differences between FORTRAN and Pascal:

- The ELSEIF and the ELSE are optional (as are their equivalents in Pascal).
- Only a single ENDIF is required in any of the variations of this construct.
- The logical expressions are enclosed with parentheses.
- The "IF (log_expr) THEN" is considered a single statement. A continuation flag must be used to extend this statement to a second line. This holds for the "ELSEIF (log_expr) THEN" statement too. This requirement will be illustrated in Section 2.3.

The IF-THEN-ELSE construction has only five variations so we can list them all. The most complex form is listed above; the other four forms follow:

1. The second form is the simple IF (no ELSEIF or ELSE):

```
          IF ( log_expr₁ ) THEN
               statement(s)

       ENDIF
```

2. The third variation has just the IF and the ELSE (no ELSEIFs). Notice that there can only be one ELSE:

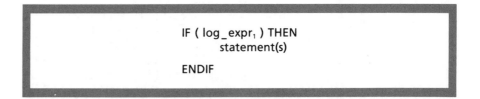

```
          IF ( log_expr₁ ) THEN
               statement(s)

       ELSE
               statement(s)

       ENDIF
```

3. The fourth form has only a single ELSEIF (no ELSE):

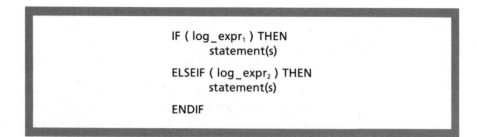

```
IF ( log_expr₁ ) THEN
        statement(s)

ELSEIF ( log_expr₂ ) THEN
        statement(s)

ENDIF
```

4. The final variation has mutliple ELSEIF (but still no ELSE):

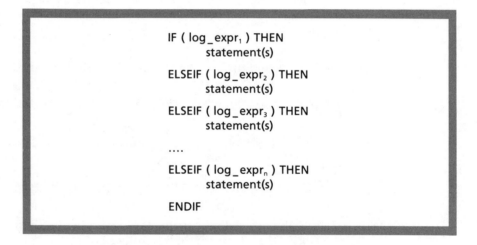

```
IF ( log_expr₁ ) THEN
        statement(s)

ELSEIF ( log_expr₂ ) THEN
        statement(s)

ELSEIF ( log_expr₃ ) THEN
        statement(s)

....

ELSEIF ( log_exprₙ ) THEN
        statement(s)

ENDIF
```

2.2 LOGICAL EXPRESSIONS

The definition of logical expression involves three points: First of all, the parentheses around the logical expression part are required. Secondly, the relational operators need to be memorized. They are quite different from their Pascal equivalents. Notice especially the use of the periods—they are required:

Operation	Meaning	Pascal
.LT.	Less Than	<
.LE.	Less than or Equal to	< =
.EQ.	EQual	=
.NE.	Not Equal	< >
.GT.	Greater Than	>
.GE.	Greater than or Equal to	> =

The final part of a logical expression refers to the logical operators. Note again the use of the period:

Operation	Meaning	Precedence	Pascal
.NOT.	Logical negation	highest	NOT
.AND.	Logical conjunction	intermediate	AND
.OR.	Logical disjunction	intermediate	OR
.NEQV.	Logical exclusive OR (NOT equivalent)	lowest	(none)
.EQV.	Logical equivalence	lowest	(none)

Here are some examples of logical expressions:

```
( INDEX .LE. 5 )
(( INDEX .GT. MSORT) .AND. (N .NE. 17 ))
```

This second example has the same meaning as the Pascal statement:

```
( INDEX > MSORT ) AND ( N < > 17 )
```

2.3 "IF-THEN" EXAMPLES

The first point I want to expand on is this: the line format is very important. Here are two versions of a trivial FORTRAN IF construct. It doesn't matter if there is one "statement" or several; the construct doesn't change.

There is no BEGIN-END in FORTRAN:

```
IF ( I .EQ. 0 ) THEN
   statement(s)
ENDIF
```

This could also be written as in the next example, but I'm sure you'll agree that this version is less readable than the previous one.

```
C2345C78901234567890123 4567890
       IF ( I .EQ. 0 )
     + THEN
                statement(s)
       ENDIF
```

The line I want you to notice is "+ THEN", the continuation of the IF construct. A "ruler" as a "comment" line was added to show you how the continuation must be entered; notice the "c" in the continuation position.

Let me try again by showing you several ways of doing the same thing in Pascal. First, assume there is only a simple "statement" in the IF (not a compound statement). Notice that the Pascal ";" behaves like the FORTRAN "ENDIF" does.

```
IF I = 0 THEN
    statement;
```

In Pascal, you could write this all on one line or on several, but there is no way to do the following in FORTRAN:

```
IF I = 0 THEN statement;
```

Now I want to make another point. Let's examine what happens in the case of a compound "statement." Pascal, of course, needs the BEGIN-END to enclose the several statements:

```
IF I = 0 THEN BEGIN
    statement;
    statement;
    ...
    statement
END
```

But FORTRAN's IF construction isn't changed at all—there are no BE-GIN-ENDs used. FORTRAN is the same whether used with one or with several statements:

```
IF (I .EQ. 0) THEN
      statement
      statement
      ...
      statement
ENDIF
```

FORTRAN IF constructs, then, are quite different from Pascal. FOR-TRAN has a single IF construction that is independent of the number of statements in it. FORTRAN doesn't need that old bug-a-boo, the semi-colon, or any BEGIN-ENDs. These features make the IF much easier to use, understand, and code in FORTRAN.

Next, we go on to investigate an IF-ELSE example. Again, I know that the semicolon and ELSE are great sources of frustration in Pascal. FOR-TRAN is much easier to understand and use:

```
IF ( I .EQ. 0 ) THEN
      FLAG = 1
ELSE
      FLAG = 0
ENDIF
```

It should be emphasized again that the following Pascal construction is *not* allowed in FORTRAN, because of the concept of one statement per line. It takes *at least* five lines to express the above construction in FOR-TRAN. Of course, the following example, or any of several other varia-tions, would be acceptable in Pascal:

```
IF I = 0 THEN FLAG = 1
ELSE FLAG = 0;
```

Finally, let's look at a nested IF in both FORTRAN and Pascal. In my experience, the use of semicolons in nested IF constructs is the hardest part of Pascal to understand. Fortunately, that problem goes away in FOR-TRAN. Here is an example; it is nonsense code and intended only to il-lustrate how to nest IF constructs. I have only nested to two levels in this example, but of course FORTRAN allows deeper nesting.

```
IF ( IN1 .EQ. 5 ) THEN
    IF ( JHEAD .NE. 100 .AND. JTAIL .GT. 0 ) THEN
        JHEAD = JHEAD + 5
    ELSE
        JTAIL = JHEAD
    ENDIF
ELSEIF ( IN1 .EQ. 7 ) THEN
    IF ( JHEAD .NE. 100 ) THEN
        JHEAD = JHEAD - 5
    ELSE
        JTAIL = JHEAD
    ENDIF
ELSEIF ( IN1 .LE. 0 ) THEN
    JHEAD = JHEAD * JTAIL
    IF ( KLAM .NE. KOUNT ) THEN
        KLAM = JHEAD
        JTAIL = KOUNT
    ELSEIF ( KLAM .LT. 0 ) THEN
        KLAM = 0
    ELSE
        KLAM = KOUNT
    ENDIF
ENDIF
```

You might have noticed by now that FORTRAN has solved the problem of the dangling ELSE. In Pascal, this problem comes up when you have an IF within another IF and only one ELSE. It drives Pascal students crazy and makes a nice test question.

```
IF X < 0 THEN
    BEGIN
    IF Y > 5 THEN
        Z := X + Y
    END
ELSE
    Z := 0
```

In the above example, the ELSE would be attached to the inner IF when the BEGIN-END pair is left out. The BEGIN-END forces the ELSE to be part of the outer IF instead. In FORTRAN this issue never comes up because every IF must have an ENDIF. Look again at the same thing in FORTRAN:

```
IF ( X .LT. 0 ) THEN
    IF ( Y .GT. 5 ) THEN
        Z = X + Y
    ENDIF
ELSE
    Z = 0
ENDIF
```

2.4 DESIGNING AND TESTING SOFTWARE

Before getting to a real, live coding example, two topics need to be discussed. While no one would disagree that "code" is important, designing and testing that code is at least as important. One major problem in writing software is the elimination of bugs. Certainly you know that bugs can be squashed during the coding phase of program development, but this section will alert you to other methods as well. Like preventive medicine, these "antibugging" methods have the goal of keeping bugs from getting into programs in the first place. Spending time designing the algorithm *before* coding is a big help. Designing tests to wring out bugs is another. This section is intended to give you some guidelines for systematically approaching these two topics.

DESIGNING

I will make the following claim: "Program development goes faster if you design your program *away* from the terminal." I don't know whether you have discovered this fact or not yet, but you should be aware that this has been proven over and over again by all sorts of studies. Let's see if the following scenario fits anybody you know.

The night before the problem is due, a student tosses his notes, text, and a can of Mellow Yellow into his bookbag, bids farewell to his video game buddies and trudges off to the terminal room. Our student logs into the system and, while the system messages are scrolling by, digs into his notes and text to find the problem set. After briefly scanning the first problem, he gets the editor to create a file for him. The problem is "easy," so coding begins immediately. "Not too shabby," our hero remarks to himself 40 minutes later, as he enters the final END. He exits the editor and starts the compiler for the first time. Forty-three errors: a pretty standard score for the first time through, nothing to be ashamed of. There are too many errors to fit on the screen at once. Only the last seven are left—and they are obscure because they are a result of other errors. No matter! Back to the editor anyhow. "I can at least hit one or two this trip." And so it goes for the next hour or so—editor, compiler, editor, compiler—gradually converging on the magic "zero errors."

When that condition finally occurs our student starts thinking that maybe he'll get to see the Wednesday Night Movie after all. Programmers are hopeless optimists: we are all guilty of believing that "the next run will be the last." Of course, the student finds error after error in his unplanned program; what looked like a piece of cake three hours ago isn't converging at all. In fact, one particular PROCEDURE is a snake's nest of flags and counters that even an expert coder couldn't understand. It's all "trial and error" coding now: change a counter to initialize at 1 instead of 0, change an "or" test to an "and" test, loop an extra time—nothing works! Going on five hours now—"got to get this hummer going!" declares the student. "Maybe a break will help."

The point is made. We've all been there—every programmer that has set fingers to keyboard, or to the keypunch for that matter. This problem has existed from the very start. In the mid-1960s, the computer industry suddenly realized how much this kind of approach was costing. Studies were conducted, theories were postulated, rejected, reformulated. Today, we state the obvious: *plan ahead*. This "solution" has many different forms but, generally, it is called "software engineering."

While this isn't a software engineering text some of the principles from that discipline have been carried into this text. We will highlight these principles right now, so that you will not only recognize them but also try them out.

The first thing you must do is to curb the urge to design at the keyboard. Instead, *make* yourself write down a step-by-step description of what your program is supposed to do. This description is called the "algorithm." Here are a few stylistic guidelines for that description:

1. Avoid computer jargon; just use English. On the other hand, don't be afraid to use words like IF, ELSE, LOOP, and UNTIL—those are English too.
2. Use mathematical notation for clarity whenever necessary.

3. Introduce English versions of variable names. If it helps, explain what you are doing. But, if variable names obscure the steps, don't use them.
4. Be sure that your description includes major loops. Indent the body of the loop to make that structure clear.
5. Number the steps, so that when you say GOTO step "x", you know what you are talking about. GOTO is English too.
6. Keep your description small, say, about 10 steps. Make use of "subdescriptions" for the complicated parts.

This list is only a set of guidelines, not absolute doctrine. You will find examples of this form throughout the remainder of the book, but don't hesitate to develop your own style. Your instructor may have different views on this subject and, when you find yourself in industry, you will probably have to follow some very definite rules. The point we want to make here is that if you haven't developed the practice of writing down an algorithm *before* entering your program, you have a very bad habit that you need to break.

There is another obvious advantage to this approach—this one is a little more pragmatic. The student in the example will eventually get his program running—all it takes is time—and *then* he will add comments. I know that and so do you. By developing the comments *first* you won't have to do such mundane things *afterward*. The algorithm you develop first is likely to be correct even after many hours spent debugging the corresponding code. Why is this? Because the algorithm is stated in general terms rather than in specifics. Chances are very good that your general statements about a problem are correct; it is the coded implementation of the algorithm that causes the several hours at the keyboard. But you must experience what we're saying for it to be convincing. If you haven't developed a program in this manner yet, try it now—you might even like it.

But, be forewarned; this is a developed discipline. Even after more than 20 years of coding, I find myself occasionally slipping back into the old bad habits. However, the method I've discovered that satisfies my tactile urges is to use the editor to jot down my algorithm. (Yes, I know I'm contradicting my original statement!) Starting at the terminal, I create a dummy PROGRAM with an END. The algorithm is inserted as comments. I may even make a data declaration or two as the need arises, and may create the subprogram names and CALLs to them as well; but that is all the code I allow myself. Then, I do the same thing with defined subroutines. When the whole thing looks complete, I start coding—one program at a time. Just to check how I'm doing, sometimes I may run the developing program through the compiler to get out the most obvious bugs right away and to give my mind a chance to rest and contemplate the next subprogram—naturally, I don't try to run the incomplete program. This method satisfies my irrational urges to "get a move on," provides some positive feedback and encouragement, while allowing me to keep my resolve not to code a program without an algorithm.

TESTING

No one is likely to argue with the claim that "testing is an art." Program "bugs" are a fact of life—even with careful designing and coding—and they must be dealt with systematically and realistically.

Just as you were encouraged to "plan" your coding by designing it ahead of time, you should also be encouraged to "plan" your testing by designing your tests before you do any coding. Is this a surprising statement? Testing is a very ego-sensitive activity; after all, programmers are reluctant to admit that *their* programs could *possibly* contain *any* errors. It is much easier to blame the computer, the instructor, the computer center staff, and the janitor than it is to blame yourself—such is the nature of humanity. But, this self-centered attitude must be overcome if you are going to write programs that are reasonably error free. The technique proposed here is widely and successfully used. However, there may be other techniques you favor.

After your design is completed and before coding begins, design your tests. By this time, you know how your program is to work but the coding details have to be worked out yet. There are several reasons for defining your test cases early:

1. Using a fresh approach to the problem could uncover design problems.
2. There is less tendency to skip "trivial" cases, since it isn't clear just what constitutes "trivial" yet.
3. There is still time to think about testing; later, certain tests may not be considered because of the pressure.
4. The programmer is still looking at the "big picture" and is better able to identify *all* the tests needed. Once coding is done, a programmer tends to be at a lower conceptual level and it is more difficult to identify meaningful tests.

Then, when testing begins you are ready with a plan to carry out. Testing can be neatly subdivided into three phases:

1. Determination of the presence of an error.
2. Location of the error in the code, and
3. Correction of the error.

In industry, the larger projects even subdivide their workforce and attack testing from these three directions. In one sense, the three are unrelated. Some people work fine in the "location" phase but are useless in the "correction" phase. And, very often, the "determination" phase is staffed by nonprogrammers.

Determination

To determine if your program has any errors, you must first know what your program is supposed to do. That may seem like a self-evident statement, but let's look at an example. Suppose you were asked to write a program to convert temperature—Celsius to Fahrenheit. What reasonable value would you pick to test your program? Does 43°C sound good? "As good as any other," you might say. "Unacceptable," I would say, because you haven't the faintest idea what the equivalent Fahrenheit temperature is. Agreed, you could get out your calculator and figure it out. But suppose you made the same error using the calculator that you made in your program? Instead, use a value you "know." What about 100°C? That makes a lot more sense, since you know what the boiling point of water is, 212°F, of course.

The point is this; you have to know the answer *before* you ask your program the question. The underlying assumption is that, if your program produces the right answer under controlled inputs, it will also work correctly when you don't know the answer. Unfortunately, you will find that this argument doesn't always hold for complex programs. In any case, the "art" of looking for an error involves selecting tests that have known answers. Sometimes, knowing the answer may require several hours worth of research—as will be illustrated in later chapters. A final caution: if you can no longer find any errors in your program, that does *not* mean that the program is error-free. Proving correctness of a program is an extremely complex process generally, and one which cannot be done for all cases—yet!

Location

So, you found that your program doesn't work correctly. What is the next step? Again, this may seem too obvious to state—but we will anyway. The next step is to locate where the error is occurring in your program. Maybe you haven't encountered this problem yet, but the student in our example was truly bogged down by this step. He didn't know where his error was; that is, he hadn't isolated the bad statement or statements. Instead, he was changing everything in sight. The "art" of locating the bad code requires entirely different skills than those required to determine if an error exists. In a complicated program, the code causing the error isn't all that obvious. Someone familiar with the program may have a general idea of where the error is, but may not necessarily know which line is causing the problem. Several techniques are worth considering at this stage:

1. You may be able to examine the code and simply "spot" the error.
2. You may look at the code, suspect a problem and invent a test case to test your theory.
3. You may have to invent some more test cases to isolate the error just to see what the failure characteristics are before even guessing at a place in the code to examine.
4. If you have absolutely no idea of the location of the error, then try to identify the parts—that is, the subprograms—that couldn't possibly cause the problem. This procedure narrows down the number of parts that may be causing trouble.

Correction

My electric furnace quit working the other day. I had no difficulties finding the problem: a fuse had blown. But simply replacing the fuse didn't fix the problem. I tried that, and the fuse blew again. I had to find *why* the fuse had blown. I had the location of the immediate failure—the fuse—but not the reason for its failure. The problem facing me was to identify what needed correcting so the fuse would not blow again.

The furnace analogy holds for software too. Once you've found an error and have determined what went wrong in your program, you don't "just" fix it. It's not that simple. In some cases the location and the correction do coincide. For instance, if one of the signs in an equation is reversed, you fix the equation. But this is not the kind of problem we need to consider. A complicated software package usually requires a complicated correction. Someone said that, in programming, "errors never disappear, they merely move around." The point is that when fixing an error, you have to be certain you're not creating a new one. Two issues need to be addressed:

1. FANIN. The line you have isolated may be executed for other input cases—and all those cases work correctly. By changing this particular line, you will invalidate those tests.
2. FANOUT. The line you have isolated may be executed to generate output for other cases. By changing this particular line, you will change the output of other tests.

When looking for a solution, you have to consider all the paths your program may take in reaching the bad line and all the paths it may take after leaving it before you make any change.

Look at the conclusion to this part of the problem: The line that caused the error may not be the line you fix at all. For example, you may have to create a new subprogram to handle this particular case, so the line that "blew up" will simply never be executed for this test.

Testing Summary

What process, then, is suggested to kill bugs? Consider the following algorithm:

1. Run the program with tests you developed before you coded, until one of them fails.
2. On the basis of the type of failure, make an assumption to explain the failure. Don't guess. Be ready to defend your position with a logical argument.
3. Locate where your program failed.
4. Determine a correction and make it.
5. Run *exactly* the same test case again.
   a) If the error still exists, "unfix" the correction and return to Step 2 because your assumption was wrong.
   b) If the error disappears, rerun all the previous tests cases to be sure you didn't introduce a new error. This is called "regression testing."
      ▪ If you get a failure during the rerunning, you have to decide if the fix you proposed should be kept or not. You may decide to keep this fix and work on the new problem or you may "unfix" this solution and return to Step 2.
6. Return to Step 1 to run a new test case.

Working on *one problem at a time* is strongly recommended. When you determine you have a bug, plunge right in and follow it through to the end. You might try writing down a statement of the bug; it can help to organize your thoughts. Then apply the recommended algorithm. It is even a good idea to keep a diary, journal, or logbook of bugs. If you encounter additional bugs along the way, don't get sidetracked; write down the second bug, but continue working on the first one. One of a student programmer's biggest difficulties is being overwhelmed by errors. Writing the second problem down ensures that it won't be forgotten; staying with the first problem ensures that your mind isn't diverted from the primary objective. When you finally have a solution for bug #1 that works, write the solution down in your logbook too. You may find that this saves you time the next time you get that kind of problem. You can also write down solutions that *don't* work; this will save you from trying the same unsuccessful change twice.

Finally, when you find you don't seem to be converging on any solution, and you seem to be embossing your fingerprints on the keys—just drop it! Start a new test case, work another problem, get a drink of water, explain your problem to somebody else—do something to get out of the psychological rut you are digging. Another "trick" you should know about is to sleep on particularly difficult bugs; your subconscious mind can do amazing things. Testing is an "art" and, to succeed, you must try every tool available.

Put It Together

For a final example, and a real program, let's look at a simplified version of the military-to-civilian time conversion problem below. This example was coded in a sloppy fashion just to illustrate the nested IF construct again. It could be coded with a single IF construction (see Problem 1) or with a symmetrical nested structure, nested for the A.M. case just like the P.M. case.

Good software engineering demands that we start with a description of the problem and an algorithm before looking at the code. A description in the form of the following table will be sufficient in this case:

Military time	Civilian time
0000	12:00 midnight A.M.
0001	12:01 A.M.
....	....
0059	12:59
0100	1:00
....	....
1159	11:59 A.M.
1200	12:00 noon P.M.
1201	12:01 P.M.
....	....
1259	12:59
1300	1:00
....	....
2359	11:59 P.M.
2400	illegal (use 0000 instead)

The corresponding algorithm is:

a) Read Military_time.
b) If 0000 <= Military_time <= 0059, then add 1200 to compute Civilian_time and it is A.M.
c) If 0100 <= Military_time <= 1159, then no computation is necessary. This is A.M. too.
d) If 1200 <= Military_time <= 1259, then no computation is necessary. This is P.M.
e) If 1300 < Military_time, then subtract 1200 to compute Civilian_time. This time is P.M. also.
f) Display Civilian_time.

The tests for this algorithm are derived from the problem description. In this case it is pretty easy to see that every illustration in the description should be tested. You may even want to add a few in between just to gain confidence in your solution.

It is usually the case that the algorithm only approximates the final program. This is true in the example that follows as well, since it was possible to combine steps "d" and "e" to make a tighter program.

```
        PROGRAM IF1
C
C Read a Military time. Convert from Military to
C Civilian time using nested IF-THEN-ELSE
C constructs.
C Print the Civilian time.
C
        WRITE (*,*) 'Enter military time:'
        READ (*,*) MTIME
        IF ( MTIME .LT. 100 ) THEN
C
C --------- Input less than 1:00 am
            MTIME = MTIME + 1200
            WRITE (*,*) MTIME, ' AM'
        ELSEIF ( MTIME .LT. 1200 ) THEN
C
C --------- Input between 1:00 am and Noon
            WRITE (*,*) MTIME, ' AM'
        ELSE
C
C --------- Input after Noon
            IF ( MTIME .GE. 1300 ) THEN
C
C ------------- Input after 1:00 pm
                MTIME = MTIME - 1200
            ENDIF
C
            WRITE (*,*) MTIME, ' PM'
        ENDIF
C
        END
```

When this program is run, the following takes place:

```
Enter military time:
1158
        1158 AM
```

Pitfalls

Since you transferred recently from Pascal, you will find that there are several inventive ways to misuse and abuse the FORTRAN IF statement. Select any one from this list:

1. Forgetting the parentheses in the IF and ELSEIF construction.

2. Putting the IF on one line and the THEN on the next without making a continuation statement.

3. Entering END rather than ENDIF.

4. Using BEGIN-END any place in the FORTRAN IF structure.

5. Using the Pascal relational operators, "<", ">", and "=" rather than the standard FORTRAN relationals.

6. Using .LS. instead of .LT. and/or .GR. instead of .GT. Trying to keep these abbreviations straight can be frustrating. Some versions of FORTRAN allow both forms of these two relationals.

Exercises

1. Modify the military conversion problem to include any or all of the following. Please note that the colon (:) is *not* to be printed, because we haven't discussed formatted output yet.
   a) Remove the nested IF test for the noon to 100 P.M. region. Make an ELSEIF clause instead.
   b) Print "Noon" (instead of 1200) for input value of 1200. Print "Midnight" (instead of 0) for input values of 0.
   c) Validate the minutes part of the input time. Print out an error message if minutes are greater than 59.
   d) Validate the hours part of the input time. Print out an error message if time is greater than 2359.

2. Create a table (as in Put It Together), write an algorithm, and then design an IF-THEN-ELSEIF program to model the following function, $f(X)$. The function has a value of zero when X is less than zero or greater than 11.

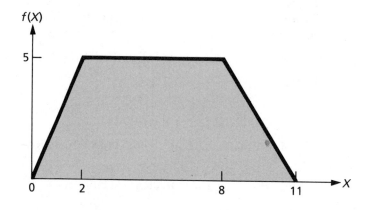

Your PROGRAM should perform the following algorithm:
a) Prompt/read a value for X.
b) Compute $f(X)$.
c) Display $f(X)$.

3. Design and code a program to model the following function:
a) Has the value $X + 10$ until it intercepts the curve X^2 from the left.
b) Has the value X^2 until it intercepts the line $-\frac{1}{2}*X + 10$ on the right.
c) Has the value $-\frac{1}{2}X + 10$ for X greater than that intersection.

Your PROGRAM should perform the following algorithm:
a) Prompt/read a value for X.
b) Compute $f(X)$.
c) Display $f(X)$.

4. Write an algorithm and then design/code a program which solves *all* the cases of the quadratic equation.
a) All combinations of real and imaginary solutions.
b) All error processing.
c) Coefficients that are nearly "equal"—that is, which cause roundoff problems.

Here are some great test cases for the equation $aX^2 + bX + c = 0$:

$$a = 3, b = -1, \quad c = -31$$
$$a = 4, b = 0, \quad c = -25$$
$$a = 1, b = 2, \quad c = 2$$
$$a = 3, b = -8, \quad c = 0$$
$$a = 1, b = -10{,}000, c = 1$$
$$a = 0, b = -10, \quad c = 2$$

You will need the square root function to solve this problem. It is used in FORTRAN the same way it is used in Pascal. Actually, the basic formula was coded for you in Chapter 1, remember?

5. Input x and y intercepts of a line. Then compute and output the slope and the Y-axis intercept of the line. This is a standard analytical geometry problem. You will find the appropriate formula in any standard textbook.

6. Design/code a program that accepts two points in space (three dimensions) and then computes and displays the distance between them. You will need your analytical geometry textbook to determine the necessary formula for the distance between two points in 3-D space.

7. Enter and run the following problem. What do you think we're trying to prove with this? What is the general form of numbers which will permit the "amazing" message to be printed?

```
      PROGRAM TRUNC
C
C Problem number 7. What's it all about?
C
      REAL X
C
      WRITE (*,*) 'Enter 999.999999'
      READ (*,*) X
      IF ( X .EQ. 999.99999 ) THEN
          WRITE (*,*) 'Amazing'
      ELSE
          WRITE (*,*) 'Just Ho Hum'
      ENDIF
C
      END
```

8. As you have been taught, $R = E/I$. But, in the real world, resistors only come in certain "standard" values. And, due to manufacturing variations, these "standard" values are allowed to vary by a certain percentage. The 20% "standard" values are:

> 68, 100, 150, 220, 330, 470, 680, 1,000, 1,500, 2,200, 3,300, 4,700, 6,800, 10,000, 15,000, 22,000, 33,000, 47,000, 68,000, 100,000, 150,000, 220,000, 330,000, 470,000, 680,000, 1,000,000, 1,500,000, 2,200,000, 3,300,000, 4,700,000, 6,800,000, and 10,000,000 ohms

You are to design/code/test a PROGRAM that will:
a) Prompt/read a resistance value.
b) Compute the closest 20% "standard" value. If the input resistance is exactly between two "standard" values, select the lower one.
c) Display that value.

This program is basically a rather long ELSEIF statement that starts like this:
1. If the input resistance is $< = 84$ (midway between 68 and 100), then Display 68.
2. Else, if the input resistance is $< = 125$ (midway between 100 and 150), then Display 100.
3. Etc.

9. Given the self-tapping screw number and the thickness of metal, the following table shows the drill size required:

Screw no.	Metal thickness	Drill no.	Screw no.	Metal thickness	Drill no.
2	16	52	4	50	41
2	20	52	4	62	39
2	25	51	4	78	38
2	31	50			
2	39	49	6	16	37
2	50	49	6	20	37
2	62	48	6	25	36
			6	31	36
4	16	44	6	39	35
4	20	44	6	50	34
4	25	43	6	62	32
4	31	42	6	78	31
4	39	42	6	99	30

You are to design/code/test a PROGRAM that will:
a) Prompt/read a screw number: 2, 4, or 6.
b) Prompt/read a metal thickness in integer mils, as the table
   indicates.
c) Compute the correct drill number.
d) Display the correct drill number.

For instance, if screw = 4 and thickness = 18 is input, then drill 44
should be displayed. If the thickness is exactly between two
entries—for instance 28 mils, and if the drills differ, then use the
smaller drill. Notice that the bigger the drill number, the smaller it
is; a number 52 is smaller than a number 48.

3

LOOPING CONSTRUCTS

There is only one looping construct in FORTRAN 77—the indexed DO—which is similar to the Pascal FOR. In FORTRAN, there is nothing like the Pascal REPEAT or WHILE, but since these are such useful constructs, we will show you how to make them.

As you probably noticed in the previous chapter, the Pascal IF is identical in function, but not in form, to the FORTRAN IF. In fact, the two are only superficially similar. In the same way, the FORTRAN DO is not simply a trivial subset or superset of the Pascal FOR; in many respects it is more powerful. Historically, the DO is the only major construct that hasn't changed throughout the FORTRAN generations. This is because it was the only "high-level" construct available in the original FORTRAN.

All other constructs were designed to translate directly and simply to machine code.

Before we introduce the DO statement, however, you need to become familiar with CONTINUE and GOTO statements.

3.1 "GOTO" AND "CONTINUE" CONSTRUCTS

Perhaps you have already been introduced to the GOTO controversy in your career, but if you haven't, be certain that the topic will come up some day. You should be warned that this is a very emotional subject among professionals—so don't be surprised when the shouting begins. Some say that GOTO should be banned from all languages, while others claim that it is necessary for clarity. We don't intend to settle the issue; GOTO is still very much a part of FORTRAN—essential, in fact, to construct REPEAT and WHILE statements as will be demonstrated. On the other hand, GOTO can easily be abused—which is what caused the controversy in the first place—so the programmer must be cautious at all times.

Here is the general form of the GOTO construct:

$$
\begin{array}{ll}
 & \text{GOTO label}_1 \\
\text{label}_2 & \text{statement(s)} \\
\text{label}_1 & \text{CONTINUE}
\end{array}
$$

where the following definitions apply:

- "label$_1$" is a statement label, one to five digits long, to which control is to be transferred;
- "statement(s)" is one or more complete FORTRAN statements that are to be "skipped" over by GOTO,
- "label$_2$" is a statement label that is used to gain access once again to "statement(s)" from some other place in the program. If the statement following GOTO has no label, there is no way to reach it; most FORTRAN compilers flag a missing label$_2$ as an error.

The meaning of the general form above is this: Upon reaching the GOTO (which can also be written GO TO), the "statement(s)" are skipped and execution picks up again at the CONTINUE statement. CONTINUE generates no executable code. Its whole purpose in life is to provide something to hang a "label" on.

Notice that the "label" on the CONTINUE has a special place in the line. This idea was introduced in Chapter 1, but this is the first time we've used it. "Labels" must be in columns one (1) through five (5). FORTRAN programs usually contain several "labels," so a good programming practice is to keep "labels" in increasing order; this will help in finding them later. The destination "label" must be on a line with a statement, not on an empty line. In my opinion, "labels" should never be attached to executable statements, even though the practice is valid. Instead, I always, put the "label" on a CONTINUE statement. I suggest that you get in that habit too. The reason is this: Programs are never static, there is always some reason to change them. If you put the "label" on a CONTINUE statement, it is trivial to insert a line after the CONTINUE. But, if you put the "label" on an executable line, you may find yourself splitting the "label" and the statement later in order to insert another line of code, and whenever you modify code in this way, mistakes can be made.

On the other hand, never use more "labels" than necessary; you will find that too many "labels" will make your program hard to read. If you have a BASIC background, be sure to understand that "labels" are not the same as BASIC statement numbers and have nothing to do with the order of statement execution as they do in BASIC.

We will introduce another use of CONTINUE in the next section, and you will be seeing GOTO throughout the rest of the book.

3.2 THE INDEXED "DO" CONSTRUCT

As mentioned earlier, the DO is the only built-in looping construct in FORTRAN. It is very similar to Pascal's FOR but is an extension of that construct. The general form of the DO statement is:

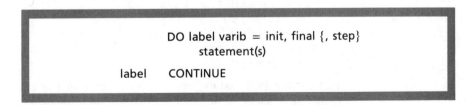

```
          DO label varib = init, final {, step}
                    statement(s)
    label     CONTINUE
```

where the following definitions apply:

- "label" is a positive integer;
- "varib" is an integer or real variable;
- "init" is the initial value assigned to "varib";
- "final" is the terminal value of "varib";
- "step" is the value added to "varib" each time through the loop and it is optional. If not specified, it is assumed to be 1. "Step" is not allowed to be zero. Note that the comma is required only when "step" is specified.
- "statement(s)" is one or more complete FORTRAN statements.

The range of DO ends at the statement (CONTINUE) marked with the "label." Notice that there is no BEGIN/END pair. With some exceptions, the "label" may be attached to any statement, but I have found that the best practice is to attach it to the CONTINUE statement. "Init," "final," and "step" can be constants, variables, or expressions. If necessary, these specifications will be converted to integers. A restriction of this construct is that the whole "DO label varib = init, final, step" statement is normally on a single line. If, because of long expressions, a second line is required, the continuation field must be set.

Let's look at a generalized example:

```
DO 100 I = 1, 10
        statement(s)
100  CONTINUE
```

This code segment will execute the "statement(s)" 10 times, just like the equivalent Pascal FOR construct:

```
FOR  I := 1 TO 10 DO BEGIN
      statement(s)
END
```

The FORTRAN DO nesting rules are the same as for the Pascal FOR construct. Here is an example of a nested DO:

```
DO 510 I = 1, 5
    DO 500 J = 4, 19
            statement(s)
500         CONTINUE
510  CONTINUE
```

Notice that two "labels" were used, and that the two are in increasing order; this is a good habit to maintain. The following Pascal code segment is like the FORTRAN code above:

```
FOR I := 1 TO 5 DO BEGIN
    FOR J := 4 TO 9 DO BEGIN
        statement(s)
    END
END
```

One more point: Nested DOs present a special case where only one CON-TINUE is required, so the FORTRAN code could also have been written as:

```
DO 500 I = 1, 5
    DO 500 J = 4, 19
        statement(s)
500 CONTINUE
```

Let's look at some other differences too. Backward loops look quite different in FORTRAN; the optional "step" is required:

```
DO 905 I = ISTRT, ISTOP, -1
        statement(s)
905 CONTINUE
```

The equivalent Pascal construct would be:

```
FOR I := ISTRT DOWNTO ISTOP DO BEGIN
    statement(s)
END
```

The "step" field constitutes the major DO/FOR loop difference: The FOR-TRAN DO variable can be changed by other than unity. For instance, in the following example, J is incremented by 5 each time through the loop: J = 1, then 6 and 11. Finally, when the loop is exited, J will be 16.

```
DO J = 1, 13, 5
    statement(s)
ENDDO
```

To do the same thing in Pascal, you need a second variable for the FOR loop.

```
J := 1;
FOR JJ := 1 TO 3 DO BEGIN
    statement(s);
    J := J + 5
END
```

Just as in Pascal, the FORTRAN DO variable cannot be changed in the loop. But, unlike Pascal, an attempt to do so will cause a compilation error. For instance, the following technique for terminating a loop prematurely will not compile:

```
DO 800 K = 10, 14
     statement(s)
     IF ( condition ) THEN
          K = 14
     ENDIF
     statement(s)
800 CONTINUE
```

Like Pascal, neither the "final" value nor the "step" value can be changed by the "statement(s)" within the DO loop range. FORTRAN computes the iteration count before the loop is started. Therefore changing "final" and/ or "step" while in the loop will not have any effect on how many times the loop is executed. The number of loop iterations is determined by the formula:

$$\text{Iterations} = \text{INTEGER part of} \, (\, (\, \text{final} - \text{init} + \text{step} \,) \, / \, \text{step} \,)$$

If the resulting iteration count is zero or negative, none of the statements in the DO range is executed. Of course the same is true of Pascal FOR loops. To illustrate the point, look at the following PROGRAM in which "step" is changed in the loop:

```
     PROGRAM DO2
     IE3 = 1
     DO 1003 I = 1, 5, IE3
          WRITE (*,*) 'I =', I
          IE3 = 2 * I
          WRITE (*,*) 'IE3 =', IE3
1003 CONTINUE
     END
```

Here is what happens. The number of iterations is computed to be 5 because IE3 is initially 1. The value of IE3 is changed in the loop, but its new value doesn't change the number of iterations, nor does it indirectly change the value of the loop variable. When run, the output of this program looks like this:

```
I =          1
IE3 =          2
I =          2
```

```
IE3 =        4
I =       3
IE3 =        6
I =       4
IE3 =        8
I =       5
IE3 =       10
```

3.3 "GOTO" FROM A "DO"

Because GOTO is such an integral part of FORTRAN, it is quite common to jump out of a DO loop before it terminates. Here are the general rules for this construct. Using GOTO in this manner is highly controversial, so we will not dwell on this point.

1. You can GOTO out of a DO loop, but not into one.
2. The exception to the previous rule is found when working with nested DO loops. You are allowed to GOTO from an inner loop to an outer one, but not from an outer loop to an inner one. For instance, the following construction (GOTO 105) is valid: it exits the DO 101 loop and continues the DO 105 loop.

```
      DO 105 I = 1, 5
            statement(s)
            DO 101 J = 2, 9
                  statement(s)
                  IF ( condition ) THEN
                        GOTO 105
                  statement(s)
101         CONTINUE
            statement(s)
105   CONTINUE
```

3.4 "REPEAT" LOOPS USING "GOTO"

Since FORTRAN doesn't contain a REPEAT construction, you will find that you will have to "manufacture" one using GOTO. Since REPEAT is such a useful construction, its general form is included for reference. However, you may find that a variation of this method is more to your taste. Here is the Pascal code to be simulated—it should look very familiar:

```
REPEAT

    statement(s)

UNTIL log_expr
```

And here is how you can do the same thing in FORTRAN. At first glance, it may look "messy," but this form is easily recognized by most programmers and is therefore quite acceptable.

```
label     CONTINUE
          statement(s)
          IF ( .NOT. log_expr ) THEN
              GOTO label
          ENDIF
```

where the following definitions apply:

- "label" is a statement label;
- "statement(s)" is one or more complete FORTRAN statements;
- "log_expr" is a logical expression.

3.5 "WHILE" LOOPS USING "GOTO"

FORTRAN 77 doesn't have a WHILE construct either—although some dialects of it have included a provision for this statement. Just as illustrated in the previous section, the WHILE can be simulated by putting IF and GOTO together. The Pascal construct we want to simulate is this:

```
WHILE log_expr DO BEGIN
        statement(s)
END
```

And here is how you can do the same thing in FORTRAN. Again, not too clean, but keep in mind that the need for both REPEAT and WHILE weren't apparent at the time that the FORTRAN language was formulated. In fact, the REPEAT/WHILE statements were an outgrowth of the GOTO controversy mentioned earlier.

```
label      CONTINUE
           IF ( log_expr ) THEN
               statement(s)
               GOTO label
           ENDIF
```

where the following definitions apply:

- "label" is a statement label;
- "statement(s)" is one or more complete FORTRAN statements;
- "log_expr" is a logical expression.

3.6 "CASE" USING "GOTO"

This is the Pascal construction that will be simulated with FORTRAN code:

```
        CASE expr OF
  val₁:    statement₁;
  val₂:    statement₂;
        . . .

  valₙ:    statementₙ
        END
```

FORTRAN code that would simulate this construct is simply an IF-ELSEIF-ELSE structure, introduced in the previous chapter. The advantage of using IF is that the condition at each step can be as complex as necessary. However, when the "val$_i$" is a simple integer, there is another form of GOTO that is helpful. This second method is presented in Appendix D.

$$\text{IF (expr = val}_1\text{) THEN}$$

$$\text{statement}_1$$

$$\text{ELSEIF (expr = val}_2\text{) THEN}$$

$$\text{statement}_2$$

$$\dots$$

$$\text{ELSEIF (expr = val}_n\text{) THEN}$$

$$\text{statement}_n$$

$$\text{ENDIF}$$

where the following definitions apply:

- "expr" is the object of the CASE search.
- "val_i" is one of the possible values of "expr."
- "$statement_i$" are the statements to be executed when "expr" is equal to "val_i."

A few Pascal dialects support a "CASE-OTHERWISE" construct. This is easily implemented in FORTRAN by adding an ELSE clause to the example above.

Put It Together

This is a good point to summarize the material presented so far. The summary is by example. Here are two programs: The first is a game, the second is a function approximation.

Example 1

Let's begin by looking in detail at a guessing game. The player is to try to guess the computer's number. To make it interesting, the computer is programmed to generate a random number. Not all systems have a random number generator available, so I have included some code to do that. You will have to trust me, unless you want to check it yourself. You have an opportunity to play with this problem in the exercises. This first example has a fabricated REPEAT in it, which begins at the

```
100   CONTINUE
```

statement right near the top. Here is the English logic of the program:

1. Initialization and game announcement.

2. Generate a random number for player to guess. Scale it to be in the range 1 to 99.

3. Read player's first guess.

4. REPEAT
   a) If guess was wrong, then print result of player's guess, either "Too big" or "Too small."
   b) Count the guess.
   c) If any guesses are left, prompt for another guess and read it.

5. UNTIL guess is correct or guess count is exceeded.

6. If guess was correct, then congratulate player.

7. If guess count ran out, print the correct number.

8. Prompt/read for another game. If so, go to step 2.

Program testing is fairly easy, if you can assume that the random number generator is correct. There are these possibilities:

1. Player makes correct guess on first try.

2. Player makes correct guess on second try.

3. Player makes correct guess on third try.

4. Player makes correct guess on fourth try.

5. Player makes correct guess on last try.

6. Player never guesses correctly.

While working on these test cases, be sure that messages such as "Too big," "Too small," or "Last guess!" all come out appropriately too.

```
          PROGRAM GAME 1
C
C GAME1 ... The computer selects a number in the range 1 - 99. The
C player is to guess in 5 tries (the house has the advantage). The
C player is cued with only two messages; too big or too small.
C
          MODULO = 2**26
          II = 123456
100       CONTINUE
          IPLAY = 0
          WRITE (*,*) 'I''ve got a number 1-99. What is it?'
C
C .. Compute a random number; the computer's number.
          II = ABS (19 * II)
          IF (II .GT. MODULO) THEN
             II = MOD ( II, MODULO )
          ENDIF
          MYNUMB = MOD ( II/3, 99 ) + 1
C
C .. Read the first guess
          READ (*,*) IGUESS
200       CONTINUE
C
C .. Print the result of the player's guess. A win is detected
C below.
          IF (IGUESS .GT. MYNUMB) THEN
             WRITE (*,*) 'Too big'
          ELSEIF (IGUESS .LT. MYNUMB) THEN
             WRITE (*,*) 'Too small'
          ENDIF
C
C .. Count the number of plays. Give him a chance to go again,
C .. but tell him on the last try.
          IPLAY = IPLAY + 1
          IF ( IGUESS .NE. MYNUMB ) THEN
             IF ( IPLAY .LT. 4 ) THEN
                WRITE (*,*) 'Guess again'
                READ (*,*) IGUESS
                GOTO 200
             ELSEIF ( IPLAY .EQ. 4) THEN
                WRITE (*,*) 'Last guess!'
                READ (*,*) IGUESS
                GOTO 200
```

```
        ELSE
            WRITE (*,*)
1              'Too many guesses. My number is:' , MYNUMB
        ENDIF
    ELSE
        WRITE (*,*) 'You win'
    ENDIF
C
C .. Another game??
    WRITE (*,*) 'Enter 1 if you want another game.'
    READ (*,*) MORE
    IF (MORE .EQ. 1) THEN
        GOTO 100
    ENDIF

    END
```

When this program is run, the following dialogue may take place:

```
I've got a number 1-99. What is it?
51
Too small
Guess again
77
Too small
Guess again
94
Too big
Guess again
81
Too small
Last guess!
88
Too big
Too many guesses. My number is:          86
Enter 1 if you want another game.
1
I've got a number 1-99. What is it?
55
Too big
Guess again
25
Too small
```

(continued)

```
Guess again
35
Too big
Guess again
32
You win
Enter 1 if you want another game.
0
```

Notice that, in the previous problem, a GOTO 200 was used to implement the "guessing" loop. As previously shown, GOTO is a necessary part of FORTRAN because the WHILE and REPEAT structures are missing entirely. But the structure coded in the game is yet another form of the loop—called the "loop and a half." Unlike WHILE, which exits from the top, and REPEAT, which exits from the bottom, this form of looping exits from the middle.

Your instructor may object to the "loop and a half" structure or may have alternative methods of implementing it. For instance, a LOGICAL variable (LPAGAN) could be introduced to create a REPEAT structure as I have illustrated in the fragment of code below. When LPAGAN (loop again) is .TRUE., the program loops to get another guess, otherwise the game is over.

The arguments behind these two coding techniques are beyond the scope of this text and, if you are interested in pursuing the GOTO controversy, you will find that there is plenty of literature to read and consider.

```
    IF ( IGUESS .NE. MYNUMB ) THEN
        IF ( IPLAY .LT. 4 ) THEN
            WRITE (*,*) 'Guess again'
            READ (*,*) IGUESS
            LPAGAN = .TRUE.
        ELSEIF ( IPLAY .EQ. 4 ) THEN
            WRITE (*,*) 'Last guess!'
            READ (*,*) IGUESS
            LPAGAN = .TRUE.
        ELSE
            WRITE (*,*)
1            'Too many guesses. My number is:', MYNUMB
            LPAGAN = .FALSE.
        ENDIF
```

```
ELSE
    WRITE (*,*) 'You win'
    LPAGAN = .FALSE.
ENDIF
IF (LPAGAN) THEN
    GOTO 200
ENDIF
```

Example 2

In this example, the irrational Napierian Base e, which has the value 2.718281828459045, will be approximated by a series expansion. The problem of representing an infinite series or a polynomial on a digital computer is very common, and you will have a chance to investigate further in the exercises that follow. In this example, three different solutions have been coded, two of them using forms of DO loops. The general formula is:

$$e = 1 + \sum_{n=1}^{\infty} \frac{1}{n!}$$

The problem is how to express this in FORTRAN; I have thought of three ways in the following algorithm:

1. Write out the entire polynomial to some limit, say n = 6, that is:

$$e = 1 + \frac{1}{1} + \frac{1}{1*2} + \frac{1}{1*2*3} + \frac{1}{1*2*3*4} + \frac{1}{1*2*3*4*5} + \frac{1}{1*2*3*4*5*6}$$

This method is very difficult to enter correctly, and generally takes several continuation lines to accomplish.

2. Set up a loop to:
   a) Evaluate the n factorial using $(n - 1)! * n$, where $(n - 1)!$ was previously computed and
   b) invert $(n - 1)! * n$; then add that term to the series.

This method doesn't require *a priori* knowledge of the final term needed but can get into numerical trouble as indicated in Exercise 7. This method is coded in an easy-to-understand DO loop.

3. Rewrite the polynomial to remove the factoral computation altogether and make nested multiplications instead: for instance, for

$n = 6$:

$$e = (((((\tfrac{1}{6} + 1) * \tfrac{1}{5} + 1) * \tfrac{1}{4} + 1) * \tfrac{1}{3} + 1) * \tfrac{1}{2} + 1) + 1$$

This method is most accurate for *any* polynomial evaluation. The DO loop required is somewhat obscure, so study it carefully.

To implement this code, as shown in the program that follows, REAL variables were used in the DO loops. This is only valid in "full" FORTRAN. You may want to consider how the loops could have been coded with INTEGER variables instead.

```
      PROGRAM MAKEE

C MAKE e ... computes the Napierian Base three ways. In all cases, only
C 10 terms are used in the series.

C  1) Compute e using brute force application of polynomial.

      Y = 2 + 1.0/2.0 + 1.0/(2.0*3.0) + 1.0/(2.0*3.0*4.0)
     1         + 1.0/(2.0*3.0*4.0*5.0) + 1.0/(2.0*3.0*4.0*5.0*6.0)
     2         + 1.0/(2.0*3.0*4.0*5.0*6.0*7.0)
     3         + 1.0/(2.0*3.0*4.0*5.0*6.0*7.0*8.0)
     4         + 1.0/(2.0*3.0*4.0*5.0*6.0*7.0*8.0*9.0)
     5         + 1.0/(2.0*3.0*4.0*5.0*6.0*7.0*8.0*9.0*10.0)
      WRITE (*,*) 'e by brute force is', Y

C  2) Compute without reevaluating n! for each term. T is (n-1)!
C     Save n! after it is computed (for the next iteration), and add its
C     inverse into the partial sum, Y.

      Y = 1.0
      T = 1.0
      DO 200 X = 1.0, 10.0
            T = T * X
            Y = Y + 1.0/T
200      CONTINUE
      WRITE (*,*) 'e without n! is', Y

C  3) Compute polynomial by nesting terms.

      Y = 0.0
      DO 300 X = 10.0, 1.0, -1.0
            Y = 1.0/X * (1.0 + Y)
```

```
300      CONTINUE
         WRITE (*,*) 'e inside out is', Y+1

   END
```

When run, the following output will be observed:

```
e by brute force is   2.718282
e without n! is       2.718282
e inside out is       2.718282
```

There doesn't seem to be any difference between the methods. Yet, as you will see, there really is. We'll return to this topic later.

Pitfalls

This section contains some of the problems you might encounter when coding your FORTRAN programs using either DO or GOTO. These are not logic problems, rather mind-set problems due to your experience with Pascal.

Problems with DO:

1. Using Pascal's FOR instead of DO and/or adding a THEN.

2. Forgetting either the "label" in the DO or the final CONTINUE statement.

3. Interchanging the "final" and the "step" values in DO.

4. Forgetting or misplacing the punctuation in DO.

5. Using ":=" instead of "=" to assign the initial value in DO.

Problems with GOTO:

1. Misconstructing WHILE or REPEAT by misusing GOTO.

2. Duplicating a "label." This is easy to do in a large program because there are usually several "labels" present—if not for GOTOs, then for DOs. A good habit is to keep labels in ascending order in your program file.

Exercises

1. Make some "spiffy" modifications to the guessing game in the example:
   a) Add more "Too big"/"Too small" messages. Select the message via another random number generation and a CASE-like construct.
   b) Add logic to tell the player if he or she is getting closer to or farther from, in absolute value, the correct number.
   c) Change the program to generate a three-digit number, and allow more guesses.

2. Design a program around the random number generation code in Example 1 above. Investigate any or all of the following:
   a) Create a new program containing just the "random number generator" code. Examine the numbers generated—that is, print the first 100 or so. Run the program a couple of times. What are your conclusions?
   b) What is the effect of changing the constants, "123456", "19", and/or "3"? This is an open-ended question; it's difficult to tell when you are done. However, an approach for study could involve making a new program that varies those constants, under programmer control, while printing the resulting random number.

3. Design a program that will add, subtract, multiply, and divide fractions used in the trades, specifically $\frac{1}{2}, \frac{1}{4}, \frac{1}{8}, \frac{1}{16}, \frac{1}{32}$, and $\frac{1}{64}$.

4. Design/code a program that will double or halve a recipe used in cooking; specifically, you must think about the units $\frac{1}{4}, \frac{1}{3}$, and $\frac{1}{2}$. But, things get a little messy when you talk about dividing $\frac{1}{4}$ cup in half, because you have to change to the tablespoon unit. Likewise, you can talk about doubling two teaspoons, which is four teaspoons rather than one tablespoon and one teaspoon. There are many special cases in this problem, so be careful.

5. Design/code a program to compute SIN using the two sums indicated below. In the first form, you can sum terms until either the nth denominator becomes too large or the nth fraction becomes too small. In the second form, you must decide a priori how many terms you will compute, since your computation proceeds from the inside set of parentheses.

a) $\text{Sin } X = X - \dfrac{X**3}{3!} + \dfrac{X**5}{5!} - \dfrac{X**7}{7!} + \ldots$

b) $\text{Sin } X = X * (1 - \dfrac{X**2}{3!} * (1 - \dfrac{X**2}{4*5} * (\ldots)))$

This series presents a special problem because of the alternating sign. You could code a variation (a) of the first form which computes all terms of like sign first, then does the subtraction. Does that variation produce a more accurate answer than version (b)? Notice that X is in radians, not degrees.

6. Design/code a program to compute $e**X$ defined below. You may use either of the techniques described in the previous problem.

$$e**X = 1 + X + \dfrac{X}{1!} + \dfrac{X**2}{2!} + \dfrac{X**3}{3!} + \ldots$$

Why doesn't this series work equally well with positive as well as negative values of X? Design/code a program that will work properly for negative X.

7. Investigate the precision of rational numbers (base 2). Look at the inverse of the positive decimal integers, "n," for increasing values of "n." Since this exercise is CPU-time consuming, limit your investigations to $\frac{1}{100}$.

a) Write a program that computes $N * (1.0 / N)$. Of course we would expect this to have the value of 1.0 for all N. But computers are finite. For what value of N does this assumption break down? How big is the error? Do all N greater than that threshold value also fail? Is the error constant?

b) Now investigate the sum of rational numbers. This time add $1.0/N$ together N times, yielding the expected value of 1.0. For what value of N does the sum fail? How big is the error? Do all N greater than that threshold value also fail? Is the error constant?

8. What does the following program do? What does EPS represent when the IF loop is done? Hint: Why does it ever stop? Code it with a WRITE in the loop to see how it behaves. Why multiply by 0.5? Try dividing by 2.0 instead and describe what happens then.

```
      PROGRAM MAKEPS
C
C Exercise number 8. What does this program do?
C
      REAL EPSP1, EPS
C
      EPSP1 = 1.5
      EPS = 0.5
1000  CONTINUE
      IF ( EPSP1 .GT. 1.0 ) THEN
            EPS = 0.5 * EPS
            EPSP1 = EPS + 1
            GOTO 1000
      ENDIF

C
      END
```

9. This is a game for you to build. What is interesting about a game is the large number of cases you have to consider—hence the IF-THEN-ELSE structures. Here is the game. It is called "eleven matches" but has many variations. Assume there are eleven matches on the table. Players take turns picking up 1, 2, or 3 matches from the pile. The player that picks up the last match is the loser. Notice that there is no tie game possible. If the computer is the first to play, it can win every time. Here is that winning algorithm:

a) The computer picks two matches.
b) The human picks "m" matches.
c) The computer picks 4 minus "m" matches.
d) Repeat the last two steps until the human loses, which will happen every time.

The complete program can be as fancy as you like, but three "levels" of sophistication could be included in your design:

a) Design/code a primary program that will win if it goes first.
b) Add logic that will win if the human goes first but makes a mistake by not playing the above algorithm. Of course, if the human goes first and plays the winning algorithm correctly, the computer will lose.
c) Add more logic to win if the human starts correctly but later in the play makes a logic error. In other words, your program

should adapt to the human's play to turn a disadvantaged position into a winner.

10. This is a really easy problem, but perhaps you should code it to be certain of the solution. The problem is included here to illustrate how an obvious formula can be wrong. Which of the following equations will produce the series: $-1, 1, -1, 1, \ldots$, as "I" increases?

a) $(-1) ** I$

b) $- 1 ** I$

Design/write a program that includes both formulas to see what happens.

11. The roots of the following polynomial are 1.1, 3.3, 4.4, 6.6, and 7.7, as you can easily verify, but that is not the problem. The problem is how to code the polynomial to give the most accurate value of Y. Hint: This problem is related to Exercises 5 and 6. Design/code a program to evaluate the following polynomial at least two different ways. Compare the results when various values of negative and positive values of X are supplied. Just how do you determine what answer is correct? Plug in $X = 7.7$. Are you able to make a general statement about the accuracy of the methods you have chosen?

$$Y = X^5 - 23.1X^4 + 199.65X^3 - 791.945X^2 + 1396.7514X - 811.69704$$

4

ADVANCED DATA TYPING AND ARRAYS

Up to now, all variables have been typed automatically by FORTRAN using the I-N rule described in Chapter 1. Now you will be introduced to many of the typing and data structure constructions. Pascal programmers will have little trouble recognizing the need for these constructions, since that is the only way Pascal operates, but you may find the FORTRAN constructions a little difficult to use initially because they are "opposite" from the Pascal construction. Another difference is that there is no TYPE statement in FORTRAN: All typing is done in the VAR-like statement. And, of course, since there is no TYPE/VAR distinction in FORTRAN, there is no need for the keywords TYPE or VAR. This chapter will show you how to assign a type (like REAL, INTEGER, etc.) to a variable and how to declare data arrays—the only FORTRAN data structure.

To put this Chapter into proper perspective, we will review the diagram presented in Chapter 1 and present details for the "declaration_section" in the following diagram:

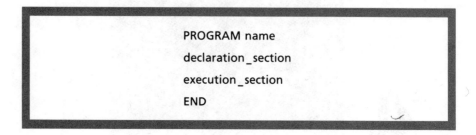

PROGRAM name

declaration_section

execution_section

END

4.1 VARIABLE DECLARATIONS

It didn't take very long for the FORTRAN inventors to realize that programmers need more control over variable types. When FORTRAN IV was introduced in 1962, a more complete data declaration mechanism was included. As Pascal programmers, FORTRAN data typing will look familiar to you, but you will find it to be "turned around": The type is first and the variable name is second. Be sure to notice that the punctuation differs from Pascal too: The colon is missing and the semicolon isn't required. The general form of the data declaration construct is:

data_type   var_list

where the following definitions apply:

- "data_type" can be any of the following keywords:
  a) CHARACTER somewhat like Pascal's CHAR, covered in Chapter 7.
  b) COMPLEX has no Pascal equivalent, covered in Chapter 8.
  c) DOUBLE PRECISION has no Pascal equivalent, it is an extension to REAL. You will be introduced to this keyword in Exercises 4 and 5.

d) INTEGER just like Pascal.

e) LOGICAL like Pascal's BOOLEAN.

f) REAL just like Pascal.

■ "var_list" is a string of one or more variable names, separated by commas. Variables declared in this way are not initialized—although some versions of FORTRAN may initialize simple variables. I usually declare only one variable in a declaration line because it is easier to include comments.

Since this concept should already be familiar to you from Pascal programming, the following example should be enough to make the transition to FORTRAN an easy one. Look at the following FORTRAN program, which contains only typing declarations:

```
PROGRAM VAR1

INTEGER INT1
INTEGER INT2, INT3
REAL REAL1
REAL REAL2, REAL3
LOGICAL LOG1, LOG2, LOG3

END
```

The first two declarations show two ways to declare INTEGERs: with a single variable in the list (INT1) and with two variables in the list (INT2 and INT3). The next two lines illustrate the same ideas using REALs. The final line shows you a LOGICAL declaration with three variables in its list. Now that is straightforward. If there is any confusion in your mind, perhaps the following Pascal counterpart will answer your questions. The program is exactly the same.

```
PROGRAM VAR1 ( INPUT, OUTPUT );

VAR
   INT1: INTEGER;
   INT2, INT3: INTEGER;
   REAL1: REAL;
   REAL2, REAL3: REAL;
   LOG1, LOG2, LOG3: BOOLEAN;

BEGIN

END.
```

Note the differences between FORTRAN and Pascal typing declarations:

1. The punctuation is different. FORTRAN is much simpler. The ";" and the ":" are not used at all in FORTRAN and the "," is used identically in both languages.
2. In FORTRAN, the keyword (INTEGER, REAL, etc.) appears first on the line, rather than last as in Pascal.
3. FORTRAN uses the keyword LOGICAL instead of Pascal's BOOLEAN.
4. The declaration section begins immediately after the PROGRAM line. There is no keyword like Pascal's VAR. More generally, there are no LABEL or TYPE declarations in FORTRAN either. A later section of this chapter will also cover FORTRAN's CONST-like construction.

4.2 I-N RULE CAUTION

As pointed out in earlier chapters, FORTRAN doesn't *require* that you declare variables, as Pascal does. Instead, you simply start using them in your program. The I-N rule is applied and the variable is automatically typed and declared. It is *never* a good practice to take advantage of this "feature" of FORTRAN and here is the reason: You may not have noticed (if you are a careful typist) that all your misspelled variables are also typed and declared automatically. This "feature" causes mysterious errors in your program. Of course, such errors become trivial typing errors when you finally discover them, but they are very difficult to detect in the first place.

Consider, for example, the number guessing program from Chapter 3. We have added a variable declaration statement and "corrected" the spelling of "IPLAY" to "PLAY" and "IGUESS" to "GUESS" to make the program more readable. However, in the process a new typo was introduced. Can you find it? FORTRAN won't! As written, the program just won't run correctly, even though it will compile without any errors.

```
      PROGRAM GAME1
C
C GAME1 ... The computer selects a number in the range 1 - 99. The
C player is to guess in 5 tries (the house has the advantage). The
C player is cued with only two messages; too big or too small. Only
C one game is played.
```

```
C
C               GUESS is the player's guess at MYNUMB
        INTEGER GUESS
C               MODULO and II are used to compute the random number
        INTEGER MODULO, II
C               MORE is a flag to indicate a new game is to be played
        INTEGER MORE
C               MYNUMB is the computer's random number
        INTEGER MYNUMB
C               PLAY is the count of guesses made so far in the game
        INTEGER PLAY
C
        MODULO = 2**26
        II = 123456
100     CONTINUE
        PLAY = 0
        WRITE (*,*) 'I''ve got a number 1-99. What is it?'
C
C .. Compute a random number; the computer's number.
        II = ABS (19 * II)
        IF (II .GT. MODULO) THEN
            II = MOD ( II, MODULO )
        ENDIF
        MYNUMB = MOD ( II/3, 99 ) + 1
C
C .. Read the first guess
        READ (*,*) GUESS
200     CONTINUE
C
C .. Print the result of the player's guess. A win is detected below.
        IF (GUESS .GT. MYNUMB) THEN
            WRITE (*,*) 'Too big'
        ELSEIF (GUESS .LT. MYNUMB) THEN
            WRITE (*,*) 'Too small'
        ENDIF
C
C .. Count the number of plays. Give him a chance to go again,
C .. but tell him on the last try.
        PLAY = PLAY + 1
        IF ( GUESS .NE. MYMUMB ) THEN
            IF ( PLAY .LT. 4 ) THEN
                WRITE (*,*) 'Guess again'
```

(continued)

```
            READ (*,*) GUESS
            GOTO 200
        ELSEIF ( PLAY .EQ. 4) THEN
            WRITE (*,*) 'Last guess!'
            READ (*,*) GUESS
            GOTO 200
        ELSE
            WRITE (*,*)
1             'Too many guesses. My number is:' , MYNUMB
        ENDIF
    ELSE
        WRITE (*,*) 'You win'
    ENDIF
C
C .. Another game ?
    WRITE (*,*) 'Enter 1 if you want another game.'
    READ (*,*) MORE
    IF (MORE .EQ. 1) THEN
        GOTO 100
    ENDIF

    END
```

What lesson can you learn from this example? Actually, there are two:

1. The FORTRAN programmer has to be aware and alert for this new way of making errors. When debugging a program, you must realize that this type of error exists.
2. There is no way to protect yourself from this type of error, unless your FORTRAN has a special extension to turn off the I-N rule.

4.3 ARRAY INTRODUCTION

Array utilization usually causes some difficulties for novice programmers, so your familiarity with Pascal is a definite advantage. You will recall that when referring to a matrix, the first (leftmost) index is the "row," the next is the "column," the third is usually called the "layer," the fourth is simply called the "fourth dimension," and so forth. The use of arrays in FORTRAN has both similarities to and differences from Pascal—as usual. Two topics must be considered:

■ array declarations and
■ subscripting.

Historically, FORTRAN has made dramatic improvements in the area of arrays. Here is what the FORTRAN I programmer had to work with:

- A special construct was required to declare an array. It was called DIMENSION and has survived to this day, even though it is no longer required.
- A three-dimensional matrix was the most complex data structure you could define in FORTRAN. Even today, FORTRAN doesn't support the concept of Pascal's RECORD.
- Subscripts could only be integer variables or integer constants, and subscript expressions were almost forbidden—rules governing their use were very complex. For instance, I*3 was valid but 3*I was not.

Most of these restrictions have been lifted in FORTRAN 77, but I'm certain there is code floating around today based on FORTRAN I rules, and someday you will probably run across a gray-haired programmer still coding to those rules—just "to be on the safe side."

4.4 ARRAY DECLARATIONS AND USE

How do you tell FORTRAN that a variable is an array structure? In the examples that follow, note a difference between FORTRAN and Pascal array usage: FORTRAN uses parentheses rather than square brackets to isolate subscripts. This is another historical footnote (as I'm sure you could have guessed), and is a result of the IBM 026 keypunch character set. The "oh-twenty-six" had no square brackets. Incidentally, it had no colon or semicolon either, which partially explains other FORTRAN/Pascal differences that have been pointed out.

The general form of an n-dimensional ARRAY declaration is:

data_type var_list (lo_index$_1$: hi_index$_1$,. . ., lo_index$_n$: hi_index$_n$)

where the following definitions apply:

- "data_type" is as previously defined;
- "var_list" is as previously defined;
- "lo_index$_i$" is a signed or unsigned integer constant and must be smaller than "hi_index$_i$".

- "hi_index$_i$" is a signed or unsigned integer constant and must be larger than "lo_index$_i$".
- "n" is the number of dimensions, and can't be larger than 7.

The conclusion is that, as expected, arrays are declared in FORTRAN somewhat as they would be in Pascal. For instance, to define a 2-dimensional matrix with two rows and five columns, a declaration would look like:

```
INTEGER MATRIX (1:2, 1:5)
```

It should be obvious that an equivalent Pascal declaration would be:

```
VAR

MATRIX: ARRAY [1..2, 1..5] OF INTEGER;
```

Of course there is more than one way to write an array declaration in both FORTRAN and Pascal. We've simply chosen the most straightforward method. Notice that the FORTRAN declaration is much simpler than Pascal. Don't be amazed at this; it simply points out the lack of FORTRAN generality. There is no way to assign the same structure to several FORTRAN matrices, short of repeating that declaration each time. For instance, consider the following Pascal code:

```
TYPE

  MATYPE = ARRAY [1..2, 1..5] OF INTEGER;

VAR

  MATRIX: MATYPE;
  OTHTIX: MATYPE;
```

It must be done in FORTRAN in the following somewhat clumsy fashion:

```
INTEGER MATRIX (1:2, 1:5)
INTEGER OTHTIX (1:2, 1:5)
```

Using a subscripted variable in FORTRAN is identical to Pascal usage, except, again, parentheses are used instead of square brackets. For instance, none of you is likely to have any problem reading the following code:

```
MATRIX (1,4) = N
MATRIX ( I, J ) = 42
MATRIX ( I, J+1) = MATRIX ( I-1, J ) + MATRIX ( I, J)
```

4.5 THE "PARAMETER" CONSTRUCT

The PARAMETER statement is similar to, but more powerful than, Pascal's CONST construction, and is intended to be used the same way. As pointed out with the IF and DO statements, this is another case where the details can't simply be extrapolated from Pascal experience. The general form of the PARAMETER statement is:

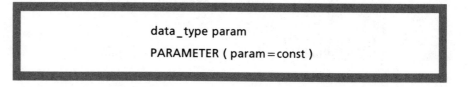

```
data_type param

PARAMETER ( param = const )
```

where the following definitions apply:

- "data_type" is as previously defined;
- "param" is the parameter's name. This name must be correctly declared prior to this statement, as we shall examine below;
- "const" is the value to be assigned to "param." It should agree in type. This may be a constant or an expression involving other "param" values, as shown in the example that follows.

Notice the use of parentheses: Although they look like excess baggage, they are required. An additional rule to keep in mind, one that was of no concern in Pascal, is that the PARAMETER, "param," should be declared like all other variables.

Here is an example program using INTEGER PARAMETERs in several different contexts. In it, four constants have been declared and PARAMETERized, then used in several expressions, and finally their final values printed. The point to be noted is that expressions involving constants and other PARAMETERs are allowed:

```
      PROGRAM PARAM
C
C Test PARAMETER construct, especially expressions...
C
      INTEGER I1
      PARAMETER (I1=3)
```

(continued)

```
INTEGER I2
PARAMETER (I2=9*14)

INTEGER I3
PARAMETER (I3=5*I1)

INTEGER I4
PARAMETER (I4=I3+I2)

WRITE (*,*) I1, I2, I3, I4

END
```

When run, the following will be displayed:

```
3    126    15    141
```

4.6 THE "DATA" CONSTRUCT

The DATA statement has no counterpart in standard Pascal. It is used in FORTRAN to initialize a variable—usually a variable that will remain constant—without using an assignment statement; for instance, it is most convenient for initialization of arrays. Furthermore, there are fancy, shorthand methods to initialize all or part of an array to the same value. Of course, the DATA statement doesn't take the place of the variable declaration itself—that has to be present too—and the constant values in the DATA statement must agree in type with that declaration.

DATA is generally a very complicated statement, and instead of showing all its forms, we will show the two most useful variations—one for simple variables and one for arrays:

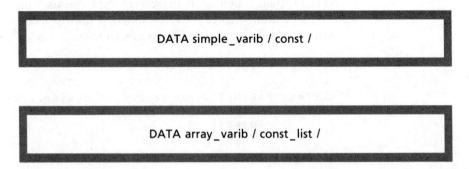

```
DATA simple_varib / const /
```

```
DATA array_varib / const_list /
```

where the following definitions apply:

- "simple_varib" is a simple variable name, one that is not an array;
- "array_varib" is a variable that has been declared as an array;
- "const" is a constant of the same type as its "simple_varib";
- "const_list" is a list of constants of the same type as the associated "array_varib."

The order of the "const_list" is important; it must be in column order. For example, the first value in the "const_list" is stored in the first element of the array, the second is stored in the second column of the first row, and so forth until every column of the first row has been filled. The first column of the second row is filled and so forth. In general, if an array is declared:

```
INTEGER A (1:n, 1:m, 1:p, 1:q)
```

then, for example, the order in which A is filled is:

$A(1,1,1,1)$, $A(1,2,1,1)$, $A(1,3,1,1)$, . . ., $A(1,m,1,1)$,
$A(2,1,1,1)$, $A(2,2,1,1)$, $A(2,3,1,1)$, . . ., $A(2,m,1,1)$,
$A(3,1,1,1)$, $A(3,2,1,1)$, . . .
. . .
$A(n,1,1,1)$, $A(n,2,1,1)$, $A(n,3,1,1)$, . . ., $A(n,m,1,1)$,
$A(1,1,2,1)$, $A(1,2,2,1)$, $A(1,3,2,1,)$, . . ., $A(1,m,2,1)$,
. . .
$A(n,1,2,1)$, $A(n,2,2,1)$, $A(n,3,2,1)$, . . ., $A(n,m,2,1)$,
. . .
$A(1,1,p,1)$, $A(1,2,p,1)$, $A(1,3,p,1)$, . . ., $A(1,m,p,1)$,
. . .
$A(n,1,p,1)$, $A(n,2,p,1)$, $A(n,3,p,1)$, . . ., $A(n,m,p,1)$,
$A(1,1,1,2)$, $A(1,2,1,2)$, . . .
. . .
$A(1,1,1,q)$, $A(1,2,1,q)$, . . .
. . .
$A(n,1,p,q)$, $A(n,2,p,q)$, $A(n,3,p,q)$, . . ., $A(n,m,p,q)$

The following example illustrates these standard DATA usages. In it, we have initialized a simple variable (PI), a 2-dimensional array (TWOD), a 3-dimensional array (THREED) and a 4-dimensional array (FOUR). A nested DO loop is included to display the second layer of THREED just to verify that everything is ok. A few additional things also need to be pointed out. Notice that there are too many digits in PI for most computers. This isn't flagged as an error, but the display of PI will be truncated to the number of digits that your computer can handle. Another point is the use of the "*" symbol in the DATA FOURD statement: It doesn't mean "multiplication," but rather "duplication." It was used to complete the array with zeros.

```
        PROGRAM DATA
C
C Demonstrate the use of the DATA statement.
C
C Declaration section:
C
        INTEGER R
        REAL PI
        INTEGER TWOD (2,3)

C
C               THREED looks like this:
C                       Layer 1:          Layer 2:
C                         x x x             x x x
C                         x x x             x x x
C                         x x x             x x x
C                         x x x             x x x

        INTEGER THREED (4,3,2)
        REAL FOURD (3,4,5,6)
C
C Initialization section:
C
        DATA PI / 3.141592653589793 /
C
C Fill TWOD with unique values:
        DATA TWOD / 11, 21,
     2              12, 22,
     3              13, 23 /
C
C Fill THREED (by columns) with unique values:
        DATA THREED / 111, 211, 311, 411,
     2                121, 221, 321, 421,
     3                131, 231, 331, 431,
     4                112, 212, 312, 412,
     5                122, 222, 322, 422,
     6                132, 232, 332, 432 /
```

```
C
C Fill the first layer of FOURD, the rest with zeros.
        DATA FOURD / 1.1, 2.1, 3.1,
     2                1.2, 2.2, 3.2,
     3                1.3, 2.3, 3.3,
     4                1.4, 2.4, 3.4,
     5          348*0.0/
C
C Execution section:
C Display the second layer of THREED in a proper Row-Column format
C
        DO 100 R=1, 4
          WRITE (*,*) THREED(R,1,2), THREED(R,2,2), THREED(R,3,2)
          WRITE (*,*)
100       CONTINUE

        END
```

This program will print the following matrix:

112	122	132
212	222	232
312	322	332
412	422	432

Because the PARAMETER and DATA "variables" are easily confused, here is a comparison of the two. First the similarities:

1. The "variable" type should be declared prior to using it.
2. The "variables" may be used on the right side of an "=" sign.

But there are more differences:

1. PARAMETERs *cannot* be changed in the program using an assignment statement—that is, they cannot be on the left side of the "=" sign.
2. PARAMETERs may be used in other declarations—anywhere a constant is valid. This is the primary purpose of PARAMETER.
3. PARAMETERs apply only to simple "variables," not to arrays.

Put It Together

In this chapter, we have presented a detailed view of the declaration portion of a FORTRAN program. What follows is a modified general format of the structure of a PROGRAM that summarizes this information:

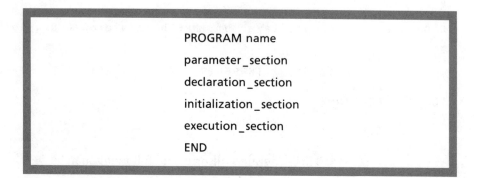

```
PROGRAM name
parameter_section
declaration_section
initialization_section
execution_section
END
```

Let us reconsider the "resistor" problem in the exercises of Chapter 2. This time, instead of using an elaborate IF structure, the resistor data will be stored in an array and searched using a DO loop. Here is the general algorithm followed in the code:

1. Initialize the "resistor" array with the DATA statement. Notice that this array begins at the low end of the range.

2. Prompt and read a resistor value.

3. Set up a loop to the end of the array.
  a) If the ith array value is larger than the input value:
    1. Decide if the ith or the $(i-1$th) resistor value should be used.
    2. Print the selected resistor value.
    3. Loop to statement 2.
  b) If the ith array value is smaller than the input value, continue the loop.

4. If the loop is exited, the last table entry should be printed as the standard value.

5. Loop to statement 2.

The testing of this algorithm should follow these lines. The sample run shows the various values used to test.

1. Carefully inspect the DATA statement for numerical types. I've used "E" notation to avoid counting zeros, thus eliminating the major source of errors in the table entry.

2. Determine if the IF-THEN-ELSE structure is correct by selecting tests at both ends and near the center of a single interval. Having tested one interval, by induction all intervals will work correctly.

3. Test both extremes of the table by entering very small and very large values that are beyond the limits of the table.

```
      PROGRAM BESTR
C
C Pick the best standard, 20% resistor from the preferred values.
C
C                          RMAX is the maximum index of the RTABLE
      INTEGER RMAX
      PARAMETER (RMAX = 33)
C                          IDXRTB is the index used for RTABLE
      INTEGER IDXRTB
C                          RMID is the midpoint between two
C                          standard values in RTABLE
      REAL RMID
C                          RTABLE contains the standard resistor
C                          values
      REAL RTABLE (RMAX)
C                          RVALUE is the user entered resistance
C                          value
      REAL RVALUE
C
C Table of R.M.A. preferred values of resistance - 20%
C Radiotron Designer's Handbook, 4th edition, Page 1337
C
      DATA RTABLE /0., 68., 100., 150., 220., 330., 470., 680.,
     1        1000., 1500., 2200., 3300., 4700., 6800.,
     2        10000., 15000., 22000., 33000., 47000., 68000.,
     3        1.0E5, 1.5E5, 2.2E5, 3.3E5, 4.7E5, 6.8E5,
     4        1.0E6, 1.5E6, 2.2E6, 3.3E6, 4.7E6, 6.8E6,
     5        1.0E7/
```
(continued)

```
C
C   Prompt/read to get actual resistance value
100      CONTINUE
         WRITE (*,*) 'Enter resistance value'
         READ (*,*) RVALUE
C
C   Loop through RTABLE to find best match
         DO 200 IDXRTB=1, RMAX
C
C   Search for table entry which is larger than entered value
            IF ( RTABLE (IDXRTB) .GT. RVALUE ) THEN
C
C   Compute the midpoint between this and previous table entry
               RMID = ( RTABLE (IDXRTB-1) + RTABLE (IDXRTB) )
     +               / 2.0
C
C   Compare entered value to midpoint, and select best value
               IF ( RVALUE .GT. RMID ) THEN
                  WRITE (*,*) RTABLE (IDXRTB), ' is standard'
               ELSE
                  WRITE (*,*) RTABLE (IDXRTB-1), ' is standard'
               ENDIF
C
C   Best resistor found and displayed - get new input
               GOTO 100
            ENDIF
C
C   Loop to get another R value from the resistor table
200      CONTINUE
C
C   Ending the loop means that entered value larger than largest
C   possible
         WRITE (*,*) RTABLE (RMAX), ' is standard'
C
C   Loop for new input
         GOTO 100

         END
```

A typical input/output dialogue may look like this. ∧C (CTRL C) is the way both MS-DOS and VAX/VMS terminate a program.

```
         Enter resistance value
         100
               100.0000000 is standard
```

```
Enter  resistance  value
125
          100.0000000  is  standard
Enter  resistance  value
126
          150.0000000  is  standard
Enter  resistance  value
150
          150.0000000  is  standard
Enter  resistance  value
−200
             .0000000  is  standard
Enter  resistance  value
1E9
     10000000.0000000  is  standard
Enter  resistance  value
^C
```

Pitfalls

When using the PARAMETER construction, be sure to declare the constant name *before* using it. If you fail to declare it, the I-N rule will apply and you may experience some very bizarre errors.

Generally, declarations will give you problems at first; the order of the declaration construction is opposite of what you learned in Pascal, and the punctuation changes a lot. If you pay attention to what you are doing, you will get it right the first time; otherwise you have to fix it. Don't be discouraged if you find yourself fixing it often; it is a common error, even for professional multilingual programmers.

When declaring and/or using arrays, you will find that you want to use the square bracket instead of the parenthesis. You will simply have to turn your thinking around—there is no place in FORTRAN for square brackets.

The biggest problem you are likely to have is the DATA statement. This construction is difficult for me to remember and do right the first time, even after years of experience with FORTRAN. I often find myself writing little programs, like the one called DATA in the previous sections, to assure myself that I am coding the DATA construct correctly. Did you notice that the name DATA was used in two different contexts in that example? It was used as a program name and in the DATA construction. Unlike Pascal, that is perfectly valid to do in FOR-TRAN, although sometimes that practice may cause confusion.

Exercises

1. Design/code a program for linear interpolation of the SIN and TAN functions. Use the following values in your program:

Deg	SIN	TAN
0.0	0.0	0.0
10	0.17365	0.17633
20	0.34202	0.36397
30	0.50000	0.57735
40	0.64279	0.83910
50	0.76604	1.1918
60	0.86603	1.7321
70	0.93969	2.7475
80	0.98481	5.6713
89	0.99985	57.290

Test your program with various values in between the values of the table. Compare the computer's results to the actual values (which you can get from your calculator or from a book of tables). Is the error constant over the range? Would it help to increase the number of significant digits in the table above?

2. This problem involves a two-dimensional linear interpolation of data. The data that follow are part of a steam table. Your program will accept temperature and pressure as input and output volume. Generally, this problem calls for three interpolations, and the order may cause different answers. We will suggest one algorithm, and

you should devise another one and compare answers. Here is an outline of our algorithm:
a) Prompt/read temperature and pressure.
b) Find the two rows on either side of the input pressure value.
c) Interpolate on both of those rows for the input temperature. This will yield two volume values, one for each of the tabular pressure entries.
d) Using the two pressure values, interpolate a third time for temperature. This will yield the desired volume.

This algorithm is only approximately correct—it doesn't take into account the fact that one or both of the input temperature or pressure values may be exactly on a row and/or column of the table. My algorithm doesn't take account of the lack of data at the lower temperatures either. Here is the steam table to use:

Pressure, lb/in²	Temperature, F						
	200	220	300	350	400	450	500
1	392.6	404.5	452.3	482.2	512.0	541.8	571.6
5	78.16	80.59	90.25	96.26	102.26	108.24	114.22
10	38.85	40.09	45.00	48.03	51.04	54.05	57.05
20			22.36	23.91	25.43	26.95	28.46
40			11.040	11.843	12.628	13.401	14.168
60			7.259	7.818	8.357	8.884	9.403
80				5.803	6.220	6.624	7.020
100				4.592	4.937	5.268	5.589

There are several parts to this problem and your instructor will decide which of them you are to do:
a) Design/code the completed algorithm as outlined above.
b) Design/code an algorithm that picks two temperature columns first, interpolates, then interpolates pressure.
c) Compare methods based on the algorithm from part (a) and the algorithm from part (b) above. Dig out more accurate steam tables to determine which of the algorithms is better over which part of the table. Why is there a difference in accuracy at all? Why isn't one algorithm better than the other all the time?

d) Linear interpolation isn't very good, actually—but it is easy. Using the bibliography in this text or your own research abilities, find a numerical analysis text book and prepare a paper that outlines an algorithm for an alternative interpolation technique.

3. This problem involves variations on the resistor problem in Pitfalls.
   a) Recode the problem using a WHILE or REPEAT structure instead of a DO loop.
   b) Run to your library and find a table of standard 5% resistor values. Modify the example in Pitfalls to operate from a 5% table instead of a 20% table.
   c) Using the information in part (b), recode the example to output both the 20% and the 5% standard values. Since the 20% values are also 5% values, the best solution to this problem would be to have only one table in it. Of course that table would have to have a flag indicating which values were only 5% standards.
   d) There are also 10% standard values. Do part (c) and output 20%, 10%, and 5% standard values. Your solution should have one table for the resistor values, and another for a flag to indicate 20/10/5% standard.
   e) The resistor data is very repetitive—identical digits in each decade. Shorten the table by taking advantage of the powers of 10 of the standard values. This problem could incorporate the concepts introduced in parts (c) and (d) above.

4. REAL numbers have a finite accuracy, as you discovered in previous problems—primarily in the Chapter 3 exercises. The accuracy can be doubled, at least, by using DOUBLE PRECISION variables instead of REAL. To demonstrate this, return to the MAKEE example in Chapter 3 and declare Y and T to be DOUBLE PRECISION, then increase the number of terms in the series from 10 to 15 or more to see how much closer the program will compute the value of "e." Compare your values to EXP(1).

5. Referring to the previous question, 1.0 is a REAL constant, while 1.0D0 is a DOUBLE PRECISION one. Does it make any difference which constant is used in the equations? Why or why not?

Matrix manipulation is an important concept in all of the physical sciences. The remainder of the problems in this chapter have to do with this process on a digital computer.

6. Design/code a program to initialize with a DATA statement, then add and print the following two matrices. You can do the calculations by hand to check your answer. This problem doesn't have any physical meaning, but it does give you an opportunity to manipulate trivial matrices.

$$
\text{MTXONE} =
\begin{bmatrix}
1.1 & 1.2 & 1.3 & 1.4 & 1.5 \\
2.1 & 2.2 & 2.3 & 2.4 & 2.5 \\
3.1 & 3.2 & 3.3 & 3.4 & 3.5 \\
4.1 & 4.2 & 4.3 & 4.4 & 4.5 \\
5.1 & 5.2 & 5.3 & 5.4 & 5.5 \\
6.1 & 6.2 & 6.3 & 6.4 & 6.5 \\
7.1 & 7.2 & 7.3 & 7.4 & 7.5
\end{bmatrix}
$$

$$
\text{MTXTWO} =
\begin{bmatrix}
0.0 & -1.0 & 2.0 & -3.0 & 4.0 \\
-4.0 & 3.0 & -2.0 & 1.0 & 0.0 \\
3.0 & -2.0 & 1.0 & 0.0 & 1.0 \\
-2.0 & 1.0 & 0.0 & 1.0 & -2.0 \\
1.0 & 0.0 & -1.0 & -2.0 & 3.0 \\
-1.1 & 2.2 & -3.3 & 4.4 & -5.5 \\
0.1 & -0.2 & 0.3 & -0.4 & 0.5
\end{bmatrix}
$$

7. Repeat Exercise 6, except subtract, rather than add, the two matrices.

8. Repeat Exercises 6 and 7 above for three 4-dimensional matrices. The order and initial condition of the matrices will be given to you by your instructor.

9. You have been introduced to determinants someplace in your education career. You may need some review, but first we have defined the notation before assigning a problem. For an order 3 matrix the determinant is:

$$
D =
\begin{vmatrix}
a_{11} & a_{12} & a_{13} \\
a_{21} & a_{22} & a_{23} \\
a_{31} & a_{32} & a_{33}
\end{vmatrix}
= a_{11}(a_{22}a_{33} - a_{23}a_{32}) - a_{12}(a_{21}a_{33} - a_{23}a_{31}) + a_{13}(a_{21}a_{32} - a_{22}a_{31})
$$

Of course you could write out this equation as a FORTRAN statement, but to get more practice with DO loops and IF statements, you are to design/code an algorithm using those constructions instead. You may use the fact that an order 2 matrix determinant is:

$$D = \begin{vmatrix} a_{11} & a_{12} \\ a_{21} & a_{22} \end{vmatrix} = a_{11}a_{22} - a_{21}a_{12}$$

Your program should read an order 3 matrix, compute the determinant, and print the result. Here are some good test data; you can check the answers by hand before running them into your program:

a) $\begin{vmatrix} 1 & 2 & 3 \\ 4 & 3 & 5 \\ 7 & 1 & 4 \end{vmatrix}$ b) $\begin{vmatrix} -3 & 1 & 2 \\ 1 & -1 & -4 \\ 2 & 3 & 5 \end{vmatrix}$ c) $\begin{vmatrix} 1 & 1 & 1 \\ 10^9 & -1 & 1 \\ 10^9 & 1 & 0 \end{vmatrix}$

10. Working with order 2 and 3 matrices is pretty much "out-of-the-book." In this problem, you are to look into the actual mechanism you are using. Look at the previous exercise to see that the order 3 determinant is derived from the order 2 definition; element a_{11} is multiplied by an order 2 submatrix. This concept is carried to higher order matrices as well. The general formula is:

$$\text{DET}(A_{1n}) = \sum_{j=1}^{n} (-1)^{1+j} a_{1j} \text{DET}(A_{1j})$$

In English, this means that the determinant is computed by summing the products of each element of the top row of the matrix with the determinant of the corresponding submatrix. The submatrix is the matrix formed without using the top row or the column of the multiplier element. The only trick in this formula is that the sign used in this summation alternates for each element: For even numbered rows it is minus, and for the odd numbered rows it is plus.

In this problem, you are to design/code a program to compute a determinant for an order 5 matrix. You should make up your own test data, but note that testing can be simplified if you keep the algorithm in mind. Notice how simple the following example

would be to work out by hand, because so many zeros have been
included:

$$
\begin{vmatrix} 1 & 0 & 0 & 0 & 0 \\ 0 & 2 & 0 & 0 & 0 \\ 1 & 1 & 1 & 2 & 3 \\ 2 & 2 & 4 & 3 & 5 \\ 3 & 3 & 7 & 1 & 4 \end{vmatrix} = 1 \begin{vmatrix} 2 & 0 & 0 & 0 \\ 1 & 1 & 2 & 3 \\ 2 & 4 & 3 & 5 \\ 3 & 7 & 1 & 4 \end{vmatrix} = 2 \begin{vmatrix} 1 & 2 & 3 \\ 4 & 3 & 5 \\ 7 & 1 & 4 \end{vmatrix}
$$

Of course this case isn't enough to flush out all of your errors, but
it will help. Remember, if you can't work out the solution by hand,
there is no way to tell if your program is correct or not.

11. What makes determinants so important? If you recall, it is one
approach to solving linear simultaneous equations—known as
Cramer's rule. A word of caution: It is not the best way, especially
for larger systems of equations, but it works for many smaller
problems. A system of 20 equations will take several thousand
years to solve! The algorithm? The research is up to you.
a) Design/code a program to read a system of 3 equations, solve
   for the 3 unknowns using Cramer's rule and print the answers.
   Here is a test case:

$$
\begin{vmatrix} 10 & -7 & 0 \\ 0 & -0.1 & 6 \\ 1 & 2.5 & 5 \end{vmatrix} \begin{vmatrix} a \\ b \\ c \end{vmatrix} = \begin{vmatrix} 7 \\ 6.7 \\ 2.3 \end{vmatrix}
$$

b) Design/code a program to read a system of 5 equations, solve
   for the 5 unknowns using Cramer's rule, and print the answers.

The remaining problems introduce you to a more common method
of solving linear systems of equations on computers. This method is
generally attributed to C.F. Gauss and it takes two steps: "elimination"
and "back substitution." The latter is easiest to understand, so it is first.

12. This problem is intended to introduce you to the "back
substitution" portion of the Gaussian elimination algorithm for
solving simultaneous equations. This algorithm will be described
by working through the example that follows. Notice how I
contrived the matrix to be almost half zeros; this is called an upper
triangular matrix. If the matrix is in this form, it is really easy (and
fast) to solve for the roots of the equations, starting at the bottom,

and working up:

$$
\begin{vmatrix} 5 & 4 & 3 & 2 & 1 \\ 0 & 5 & 4 & 3 & 2 \\ 0 & 0 & 5 & 4 & 3 \\ 0 & 0 & 0 & 5 & 4 \\ 0 & 0 & 0 & 0 & 5 \end{vmatrix} \begin{vmatrix} a \\ b \\ c \\ d \\ e \end{vmatrix} = \begin{vmatrix} -5 \\ -4 \\ -3 \\ -2 \\ -1 \end{vmatrix}
$$

The last row represents the equation:

$$5e = -1$$
$$e = -0.2$$

which, of course, is one of the five solutions. The next-to-the-last equation looks like this:

$$5d + 4e = -2$$

and since we already have a value for "e," it can be substituted into this equation and "d" can be determined:

$$5d + 4(-0.2) = -2$$
$$d = (-2 - 4(-0.2))/5$$
$$d = (-2 - (-0.8))/5$$
$$d = -1.2/5$$
$$d = -0.24$$

and so forth, for each row of the matrix, to get all of the solutions. You are to design/code a program that will read the 5 equations of an upper triangular matrix, back substitute and print the values of 5 unknowns. Compare the running time of this program to the program you designed for Exercise 11(b). Of course it is a special case, but you will be surprised at the savings!

13. This problem is designed to introduce you to the "forward elimination" part of the Gaussian elimination algorithm. Just as in Exercise 12, we will show you the algorithm by way of an example; you must synthesize the algorithm before you start programming. Here is the matrix to be "triangularized":

$$
\begin{vmatrix} 4 & 1 & 7 \\ -2 & 3 & 6 \\ 2 & -6.5 & 2 \end{vmatrix} \begin{vmatrix} x \\ y \\ z \end{vmatrix} = \begin{vmatrix} -8 \\ 0 \\ 19 \end{vmatrix}
$$

In describing what is happening, we will use the notation of Exercise 9. The first step is to make a_{21} zero. This is done by modifying the first equation and then adding it to the second. The multiplying factor is $-(-2/4)$, that is, $-(a_{21}/a_{11})$. Let's see how that works: The first equation becomes (temporarily):

$$-(-2/4)4x + (-(-2/4)1y) + (-(-2/4)\,7z) = -(-2/4)(-8)$$

which, when the math is performed, becomes:

$$2x + 0.5y + 3.5z = -4$$

Now, the first two equations are added together:

$$
\begin{array}{l}
2x + 0.5y + 3.5z = -4 \\
-2x + 3y + 6z = 0 \\
\hline
0x + 3.5y + 9.5z = -4
\end{array}
$$

See what happened: The a_{21} term was eliminated and the second row of the matrix is changed. At this point, the system of equations has become:

$$
\begin{vmatrix} 4 & 1 & 7 \\ 0 & 3.5 & 9.5 \\ 2 & -6.5 & 2 \end{vmatrix}
\begin{vmatrix} x \\ y \\ z \end{vmatrix}
=
\begin{vmatrix} -8 \\ -4 \\ 19 \end{vmatrix}
$$

Now, let's do the same thing with the first and the last row. This time the multiplier is $-(a_{31}/a_{11})$ or $-(1/2)$. The coefficients of the first row are multiplied, and the two equations to be added are:

$$
\begin{array}{l}
-2x - 0.5y - 3.5z = 4 \\
2x - 6.5y + 2z = 19 \\
\hline
0x - 7y - 1.5z = 23
\end{array}
$$

And the resulting system of equations becomes:

$$
\begin{vmatrix} 4 & 1 & 7 \\ 0 & 3.5 & 9.5 \\ 0 & -7 & -1.5 \end{vmatrix}
\begin{vmatrix} x \\ y \\ z \end{vmatrix}
=
\begin{vmatrix} -8 \\ -4 \\ 23 \end{vmatrix}
$$

Examine what has happened now. The first column is all zeros, except for the first element. The next step is to make the second column all zeros except for the top two elements. This is done by multiplying the second equation by $-(a_{32}/a_{22})$ and adding it to the third equation. This results in the following system of equations.

$$
\begin{vmatrix} 4 & 1 & 7 \\ 0 & 3.5 & 9.5 \\ 0 & 0 & 17.5 \end{vmatrix}
\begin{vmatrix} x \\ y \\ z \end{vmatrix}
=
\begin{vmatrix} -8 \\ -4 \\ 15 \end{vmatrix}
$$

The task is accomplished; the matrix has been triangularized. This completes the "elimination" step and if you take the time to "back substitute," you will find that $x = -2.6326531$, $y = -3.4693878$, and $z = 0.8571429$.

You are to design/code a program that will read 5 equations, forward eliminate, and display the resulting upper triangular matrix. This isn't the end of the story of course, and the next problem shows how to put the two parts (Exercises 12 and 13) together to create a program that quickly solves linear simultaneous equations.

14. Program the Gaussian elimination algorithm for solving an nth order linear system of simultaneous equations by combining the programs derived in Exercises 12 and 13. Gather the code from the previous two problems into a single program, and modify that code to accept any order up to 10. If you didn't do 12 and 13, this is an opportunity to "divide and conquer" rather than to try the whole thing at once; do one part at a time—design/code/test—then put the two pieces together. The correct algorithm should be obvious:
   a) Prompt/read for "n," the order of the system of equations.
   b) Prompt/read "$n+1$" values (one equation) "n" times.
   c) Display the system of equations for reference.
   d) Generate the upper triangular matrix using the elimination method outlined in Exercise 13.
   e) Back substitute using the method outlined in Exercise 12.
   f) Print the "n" solutions.
   g) Loop back to step (a).

15. Caution: Even though the Gaussian method will solve a wide variety of simultaneous equations, there are classes of equations on which it doesn't work well. Here is an example:

$$\begin{vmatrix} 10 & -7 & 0 \\ -3 & 2.099 & 6 \\ 5 & -1 & 5 \end{vmatrix} \begin{vmatrix} a \\ b \\ c \end{vmatrix} = \begin{vmatrix} 7 \\ 3.901 \\ 6 \end{vmatrix}$$

The solutions should be $a=0$, $b=1$, and $c=1$. How does that solution compare with the solution you get on your machine? What went wrong? What do you think characterizes a "difficult" system of equations? You will probably have to work through the example above by hand to see the answer. How can the Gaussian elimination algorithm be corrected to detect and compensate for this difficulty?

Research this problem and present your discoveries to the class. If you have a good library on your system, try the above example on all simultaneous equations solvers that are available, and present the various solutions. It turns out that this is a well understood problem with a rather elaborate solution. Don't bother to code a solution, it has been done.

In the future, if you ever need to solve linear simultaneous equations, look for the latest, most glamorous program in your library and then try the "trivial" problem we've just discussed on it, just to see how good it really is.

16. Using the program developed in Exercise 14, solve a 2-D truss problem.

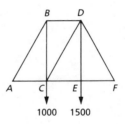

17. Using the program developed in Exercise 14, solve the following circuit problem.

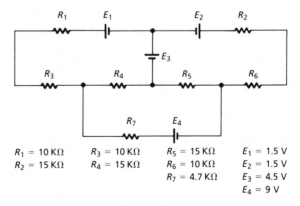

$R_1 = 10\,K\Omega$	$R_3 = 10\,K\Omega$	$R_5 = 15\,K\Omega$	$E_1 = 1.5\,V$
$R_2 = 15\,K\Omega$	$R_4 = 15\,K\Omega$	$R_6 = 10\,K\Omega$	$E_2 = 1.5\,V$
		$R_7 = 4.7\,K\Omega$	$E_3 = 4.5\,V$
			$E_4 = 9\,V$

5

SUBPROGRAMS

We are making the assumption that you already know about subprograms from your Pascal days. You undoubtedly know that the best way of designing software is by fragmenting your problem into smaller and smaller parts, until you have "manageable" units. That concept has been around for a long time in FORTRAN too: since FORTRAN I, in fact, the very first release—although it wasn't documented until later. Very few changes have been made to the subprogram constructions over the years.

In order to accommodate subprograms, the outline of a FORTRAN program has to be modified once again; detail must be added to the "execution_section." The following diagram will give you an overview of what we are about to do. Notice, in particular, the position of the main program's END statement.

```
                    PROGRAM name

                    parameter_section

                    declaration_section

                    initialization_section

                    main_program_section

                    END

                    subprogram_section
```

This chapter will concentrate on the "subprogram_section." With one exception, the "subprogram_section" is a repeat of the above figure: The "subprogram_section" is not nested (see the next diagram). The "subprogram_section" can be repeated as many times as necessary, as long as it follows the main program END statement. Here is what we will be concentrating on for the remainder of the chapter:

```
                    subprogram_declaration

                    argument_declaration_section

                    parameter_section

                    declaration_section

                    initialization_section

                    execution_section

                    END
```

As we examine subprograms—called FUNCTIONs and SUBROUTINEs—you will notice that the two are closely related, as you probably expect from your Pascal experience. Unlike the previous chapters, examples will be included as we go along. Once you know how to write your own subprograms, later in the chapter you will be introduced to the standard library of subprograms supported by FORTRAN 77.

As we run through this section, I would like you to be especially alert for the introduction of the following concepts—the little things that make FORTRAN subprograms different from Pascal:

- Like Pascal, FORTRAN supports two types of subprograms, called SUBROUTINE and FUNCTION.
- *All* arguments are Pascal VAR-type, that is, they are both input and output. This is the only type of argument that FORTRAN supports.
- The dummy arguments aren't declared in the argument list as they are in Pascal. Instead, they are declared, just like any other variable, in what I have called the "argument_declaration_section."
- FUNCTIONs are invoked in the same way they are executed in Pascal, but SUBROUTINEs are not. In FORTRAN they require the keyword CALL to be added.
- Subprogram definitions (instructions) *follow* the main program. Subprograms can't be nested, as they can in Pascal, but they may be in any order; generally we use alphabetical ordering to permit them to be found more easily.
- Subprograms may invoke or CALL other subprograms just as in Pascal. But recursion isn't supported in FORTRAN, so subprograms can't call themselves. Generally speaking, recursion can also be performed using a stack. However, this technique is not often used in FORTRAN because it is prone to errors and misunderstanding.
- All data used by a subprogram must be passed as an argument; that is, Pascal's variable scoping rules don't apply in FORTRAN. In particular, variables declared in the main program are *unavailable* to subprograms. Every rule has its exception, and we will study this exception in Chapter 9.
- The main program and all subprograms that call a user-defined function must *also* declare the function's type. Failure to do so will result in the I-N rule being invoked. FORTRAN thinks that the called function is just another "variable"—an array, in fact—if you don't tell it otherwise.
- FUNCTIONs that do WRITEing cannot be invoked in an I/O list. This amounts to WRITE recursion, and recursion is not permitted in FORTRAN.

5.1 THE "FUNCTION" CONSTRUCT

The general form of the FUNCTION declaration follows. Note the order of the keywords in this construct; it is very different from Pascal:

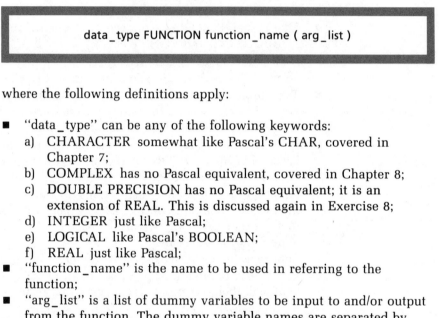

data_type FUNCTION function_name (arg_list)

where the following definitions apply:

- "data_type" can be any of the following keywords:
  a) CHARACTER somewhat like Pascal's CHAR, covered in Chapter 7;
  b) COMPLEX has no Pascal equivalent, covered in Chapter 8;
  c) DOUBLE PRECISION has no Pascal equivalent; it is an extension of REAL. This is discussed again in Exercise 8;
  d) INTEGER just like Pascal;
  e) LOGICAL like Pascal's BOOLEAN;
  f) REAL just like Pascal;
- "function_name" is the name to be used in referring to the function;
- "arg_list" is a list of dummy variables to be input to and/or output from the function. The dummy variable names are separated by commas. The dummy variables should be declared, and I suggest that this be done immediately following the FUNCTION statement. If these variables are not declared, the I-N rule is applied.

As in Pascal, within the FUNCTION executable code you must tell FORTRAN what value to return as the FUNCTION's value. This is done with an assignment statement just as in Pascal: The "function_name" is used on the left side of the assignment statement. It is well to point out again that no recursion is permitted in FORTRAN. This has a benefit: The "function_name" may be used within the FUNCTION like any other variable would be used (see the example that follows, such as TRAP = TRAP + . . .).

When a FUNCTION is to be invoked (or called), the FUNCTION's name is usually on the right side of an assignment statement, within a formula, but this is not the only context in which a FUNCTION can be used. It can also be an element of a WRITE I/O list, in an IF expression, or in any of the elements of a DO construct. When invoked, the following general form applies:

function_name (actual_list)

where the following definitions apply:

- "function_name" is the name defined in the FUNCTION statement;
- "actual_list" is the list of actual variables and constants that will replace the dummy variables in the "arg_list." The actuals are separated by commas. The members of the "actual_list" match up one-to-one with the members of the "arg_list" just as they do in Pascal.

You are responsible for making sure the order of the variables in the "actual_list" is the same as in the "arg_list," because FORTRAN makes no checks. You also have to be sure that you provide the correct number of arguments to the FUNCTION, since there are no checks for that either. These points are made in an example in the Pitfalls section of the chapter.

One last, very important point: The calling subprogram (or main program) must declare the function in the following manner:

data_type function_name

If not declared, FORTRAN will think "function_name" is an undeclared array.

5.2 "FUNCTION" EXAMPLE

In this example, we have coded a program that integrates a function using the trapezoidal rule. You may remember from your introductory integration days that this is only an approximation to the actual solution. But that is the way integration in general is done: Analytical solutions work only in specific cases. You may also remember that other approximations can be used for integration. In Exercise 4 you are asked to create a program to integrate using the rectangle rule. Neither of these methods is the last word in numerical integration. Much better techniques are available, and should be used professionally. However, this is not a numerical analysis course, and both the trapezoidal and rectangular methods are easy to code and understand.

A few additional items need to be pointed out before getting started with this example. First, it is coded so you can easily change the function that is being integrated. Exercise 3 makes particular use of this design feature. Second, you may be puzzled when you first start working with the algorithm that follows—but there is a reason for the way it is presented, and when you do Exercise 1, you will understand. Finally, the purpose of the loop in the main program is to illustrate more clearly how, by increasing the number of trapezoids, the approximation reaches the exact value. The loop also underscores the fact that this is only an approximation, not an exact solution. Notice that more (or smaller) intervals don't necessarily make a more accurate answer. This is not a failure of the theory, but another illustration of the effects of computer-introduced round-off errors. Here is a general form of the algorithm we followed:

MAIN PROGRAM

1. Prompt for the lower and upper bounds of the integral and read those two values.
2. Initialize the number of trapezoids in the approximation.
3. Set up a loop for 14 iterations:
   a) Display the number of trapezoids in the current iteration.
   b) Compute the length of the base of the trapezoid.
   c) Using TRAP, compute and sum the area of each trapezoid.
   d) Display the integral's value.

FUNCTION X

1. Evaluate the dependent variable.
2. Return that value.

FUNCTION TRAP

1. Initialize the value of the integral and of the independent variable.
2. Set up a loop for the number of trapezoids:
   a) Compute the function F at the two sides of the trapezoids.
   b) Compute the area of the trapezoid.
   c) Accumulate the area.
3. Return the final area—that is the sum of all the trapezoids' areas.

Before you look at the code there are several points I want to under-line. The first point is just to dazzle your friends: "Quadrature" is the term used by numerical analysts for numerical approximation of definite integrals, and that is exactly what I have done. This distinction is made to avoid confusion with numerical integration of ordinary differential equations.

I have imposed my "style" of program format in the following code, and you may follow it or not, depending on you and your instructor. My style is arbitrary for the most part but as you will see, it can be justified. The key points of this style are:

1. Comments are associated with each declaration. These comments are indented to keep the declaration itself visible, thus making it easy to locate.
2. Dummy argument declarations are separated for the variable declarations. The argument's usage—input, output or input/output— is indicated in the body of its comment.
3. Subprogram and main program variables are declared in alphabetical order, too, and follow the argument declarations.
4. Subprograms are delineated with a line of stars to make them easier to locate.
5. Nested statements are indented.
6. There is lots of spacing between the different parts of each statement. FORTRAN doesn't care about spaces, but my computer's text editor does. It uses a single keystroke to advance to the next word as well as a character-by-character advance. Spaces make it easier to move through the program when I have to correct it.
7. Comments are liberally salted throughout the program.

Concerning FORTRAN details, here are some points you should be especially aware of:

1. The main program calls TRAP, and so TRAP is declared just like a variable would be. This same comment applies to F's declaration in the TRAP subprogram.
2. Looping 14 times was based on the fact that 2^{14} is a reasonable upper bound on my machine; it takes about 5 seconds to compute that last integral. If I doubled it once more, I'd have to wait longer than I like. Based on execution speed of your machine, you may want to change the upper limit.
3. I've used the same name in the dummy arguments that I used when I called the function. This is not a FORTRAN requirement, but it helps me make sure that the arguments are used in the same order in which they were declared.

```
      PROGRAM QUAD
*
* Quadrature computations (using FUNCTION subprograms);
* approximation method using the trapezoidal method.
* The integral is:
*
*     /XUPPER
*    /
*   | x**3 dx
*    /
*   / XLOWER
*
*                        AREA is the value of the integral
      REAL AREA
*                        COUNT keeps track of how many times STEP
*                        has been doubled.
      INTEGER COUNT
*                        F is the name of the Function which
*                        evaluates X-cubed - the function to be
*                        integrated.
      REAL F
*                        STEPS is the number of trapezoids the
*                        interval is broken into
      INTEGER STEPS
*                        TRAP is the name of the Function which
*                        computes the area of the trapezoid
      REAL TRAP
*                        XINC is the X-axis dimension of the
*                        trapezoid
      REAL XINC
```

```
*                       XLOWER & XUPPER are the lower and
*                       upper bounds, on the X-axis, of the
*                       integral
        REAL XLOWER, XUPPER
*
* Begin Main Program...
*
* Prompt for the lower and upper bounds:
*
        WRITE (*,*) 'Input X lower and upper bounds'
        READ (*,*) XLOWER, XUPPER
*
* Initialize to 2 trapezoids
*
        STEPS = 2
*
* Setup loop - compute area (integral) based on STEPS trapezoids
* each time through
*
        DO 100 COUNT = 1, 14
*
* Display the current value of STEPS
*
            WRITE (*,*) STEPS, ' intervals.'
*
* Compute the size of the base of the trapezoid
*
            XINC = ( XUPPER - XLOWER ) / STEPS
*
* Evaluate total area using trapezoidal rule. Display the result.
*
            AREA = TRAP ( XINC, XLOWER, XUPPER, STEPS )
            WRITE (*,*) AREA, ' by trapezoidal rule.'
*
* Double the number of trapezoids
*
            STEPS = STEPS * 2
*
100     CONTINUE
*
        END
```

(continued)

```
**********************************************************************
* This is a definition of the function to integrate - X**3
* To integrate a different function, this is the only thing that has
* to be modified.
*
       REAL FUNCTION F ( X )
*
*                        X (input) is the dependent variable of the
*                        function
       REAL X
*
* Function evaluation
*
       F = X**3
*
* End of FUNCTION F
       END
**********************************************************************
* Evaluate using trapezoidal rule:
*
       REAL FUNCTION TRAP ( XINC, XLOWER, XUPPER, STEPS )
*
* ARGUMENTS *************
*                        STEPS ( input ) is the number of trapezoids
*                        the interval is broken into
       INTEGER STEPS
*                        XINC (input) is the X-axis dimension of
*                        the trapezoid
       REAL XINC
*                        XLOWER (input) & XUPPER (input) are the
*                        lower and upper bounds, on the X-axis, of
*                        the integral
       REAL XLOWER, XUPPER
*
* VARIABLES *************
*                        F is the name of the Function which
*                        evaluates X-cubed - the function to be
*                        integrated.
       REAL F
*
       INTEGER N
*                        X is the independent variable of the
*                        function
       REAL X
```

```
*
* Initialize the value of the function, TRAP and
* the independent variable, X
*
        TRAP = XINC / 2 * (( F(XLOWER)) + ( F(XUPPER)))
        X = XLOWER
*
* Set up a loop to compute the area of each trapezoid using
* F and sum those areas
*
        DO 100 N = 2, STEPS
            X = X + XINC
            TRAP = TRAP + XINC * F ( X )
100     CONTINUE
*
* End of FUNCTION TRAP
        END
```

This produces the following output for bounds of 0–2. Of course, by an-alytical methods the correct answer is 4. Look at the output to see how the trapezoidal rule converges on that solution—which, of course, is ex-actly what the theory predicts. Testing of this "system" is quite different from previous programs. The difficulty lies in the fact that the "by hand" solution isn't particularly easy to do—especially with, say, 128 trapezoids. The following tests could be considered:

1. Test to see if the "bounds" are properly read by examining the converging results.
2. Work the problem on paper for "a few" cases, perhaps 2 and 4 intervals, and confirm those results.
3. Make sure that the intended number of intervals, in this case 2^{14} intervals, are computed.

The more important impact of this program is that a function with no analytical solution will "work" too (with certain restrictions). For in-stance any complicated discontinuous function could be plugged into FUNCTION F as you will see in Exercises 5 and 6.

Output for this program will look like this:

```
Input X lower and upper bounds
            2 intervals.
        5.0000000 by trapezoidal rule.
            4 intervals.
```

(continued)

```
4.2500000 by trapezoidal rule.
   8 intervals.
4.0625000 by trapezoidal rule.
  16 intervals.
4.0156250 by trapezoidal rule.
  32 intervals.
4.0039060 by trapezoidal rule.
  64 intervals.
4.0009770 by trapezoidal rule.
 128 intervals.
4.0002440 by trapezoidal rule.
 256 intervals.
4.0000610 by trapezoidal rule.
 512 intervals.
4.0000160 by trapezoidal rule.
1024 intervals.
4.0000030 by trapezoidal rule.
2048 intervals.
4.0000000 by trapezoidal rule.
4096 intervals.
4.0000030 by trapezoidal rule.
8192 intervals.
3.9999990 by trapezoidal rule.
16384 intervals.
4.0000020 by trapezoidal rule.
```

5.3 THE "SUBROUTINE" CONSTRUCT

Let's proceed to the SUBROUTINE declaration. The general format is:

> SUBROUTINE subroutine_name (arg_list)

where the following definitions apply:

- "subroutine_name" is the name of the subroutine, to be used when invoking it;
- "arg_list" is a list of dummy variables to be input to, and/or output from, the subroutine. The dummy variable names are separated by

commas. The dummy variables should be declared. It is suggested that these declarations immediately follow the SUBROUTINE statement. If the arguments aren't declared, the I-N rule is applied to them.

When the SUBROUTINE is invoked (or called), the following general format applies. This is different from Pascal:

CALL subroutine_name (actual_list)

where the following definitions apply:

- "subroutine_name" is the name defined by the SUBROUTINE statement;
- "actual_list" is the list of actual variables and constants to be used by the SUBROUTINE. The actuals are separated by commas and, like Pascal, are matched one-to-one with the "arg_list" definition.

5.4 "SUBROUTINE" EXAMPLE

I have coded exactly the same problem again (see the previous example), this time using SUBROUTINEs instead of FUNCTIONs. The differences are subtle: in Exercise 2 you are asked to list those differences. When the program is run, the results are identical to the FUNCTION example, as you would expect. The algorithm is the same too, of course, so we won't repeat it. However, I have added comments to the code to make it a good, complete program.

Some differences need to be underscored:

1. FUNCTION F has been replaced by SUBROUTINE SUBF to reduce confusion. Notice that SUBF doesn't have to be declared in the calling program—the SBTRAP in this case.
2. SBQUAD and SBTRAP are necessarily abbreviated because of the 6-character variable name restriction in FORTRAN.
3. Output arguments have been added to the argument lists on the right end so that these SUBROUTINEs more closely resemble the FUNCTIONs in the previous example.

```
      PROGRAM SBQUAD
*
* Quadrature computations (using SUBROUTINE subprograms):
* approximation method using the trapezoidal method.
* The integral is:
*
*     /XUPPER
*    /
*   |   x**3 dx
*    /
*   / XLOWER
*                          AREA is the value of the integral
      REAL AREA
*                          COUNT keeps track of how many times STEP
*                          has been doubled.
      INTEGER COUNT
*                          STEPS is the number of trapezoids the
*                          interval is broken into
      INTEGER STEPS
*                          XINC is the X-axis dimension of the
*                          trapezoid
      REAL XINC
*                          XLOWER & XUPPER are the lower and upper
*                          bounds, on the X-axis, of the integral
      REAL XLOWER, XUPPER
*
* Begin Main Program...
* Prompt for the lower and upper bounds:
      WRITE (*,*) 'Input X lower and upper bounds'
      READ (*,*) XLOWER, XUPPER
*
* Initialize to 2 trapezoids
      STEPS = 2
*
* Setup loop - compute area (integral) based on STEPS trapezoids
* each time through
      DO 100 COUNT = 1, 14
*
* Display the current value of STEPS
         WRITE (*,*) STEPS, ' intervals.'
```

```
* Compute the size of the base of the trapezoid
         XINC = ( XUPPER - XLOWER ) / STEPS
*
* Evaluate total area using trapezoidal rule. Display the result.
         CALL SBTRAP ( XINC, XLOWER, XUPPER, STEPS, AREA )
         WRITE (*,*) AREA, ' by trapezoidal rule.'
*
* Double the number of trapezoids
         STEPS = STEPS * 2
*
100      CONTINUE
*
         END
***********************************************************************
* This is a definition of the function to integrate - X**3
* To integrate a different function, this is the only thing that
* has to be modified.
         SUBROUTINE SUBF ( X,F )
*
* ARGUMENTS *************
*                         F (output) is the value of the
*                         function at X
         REAL F
*                         X (input) is the dependent variable of
*                         the function
         REAL X
*
* Function evaluation
         F = X**3
*
* End of SUBROUTINE SUBF
         END
***********************************************************************
* Evaluate using trapezoidal rule:
         SUBROUTINE SBTRAP ( XINC, XLOWER, XUPPER, STEPS, TRAP )
*
* ARGUMENTS *************
*                         STEPS (input) is the number of trapezoids
*                         the interval is broken into
         INTEGER STEPS
```

(continued)

```
*                         TRAP (output) is the value of integral
*                         of the function
          REAL TRAP
*                         XINC (input) is the X-axis dimension
*                         of the trapezoid
          REAL XINC
*                         XLOWER (input) & XUPPER (input) are the
*                         lower and upper bounds, on the X-axis, of
*                         the integral
          REAL XLOWER, XUPPER
*
* VARIABLES **************
*                         F is the value of the function at X
          REAL F
*                         FLO and FUP contain the value of the
*                         function at the two endpoints
          REAL FLO, FUP
*                         N is a counter in the loop
          INTEGER N
*                         X is the independent variable in the
*                         computations
          REAL X
*
* Initialize the value of the function, TRAP and
* the independent variable, X
          CALL SUBF ( XLOWER, FLO )
          CALL SUBF ( XUPPER, FUP )
          TRAP = XINC / 2 * ( FLO + FUP )
          X = XLOWER
*
* Set up a loop to compute the area of each trapezoid using
* F and sum those areas
          DO 100 N = 2, STEPS
               X = X + XINC
               CALL SUBF ( X, F )
               TRAP = TRAP + XINC * F
100       CONTINUE
*
* End of FUNCTION TRAP
          END
```

The output from this program is identical to the FUNCTION example, so we won't repeat it.

5.5 BUILT–IN SUBPROGRAMS

FORTRAN has many built-in, or "instrinsic," functions. They are built-in in the sense that you don't have to declare them to use them, and you can use them with various data types. For instance, ABS (absolute value) can be used with either a REAL or an INTEGER input variable, and the FUNCTION is automatically "declared" to agree in type with the input, so that you get an INTEGER output when you use an INTEGER as input. Thus, ABS is called a "generic" intrinsic because the proper ABS function is selected by the compiler for the context of its usage.

As an introduction, we have included a partial table of generic intrinsic FUNCTIONs. The complete table is included in Appendix A.

To read the table that follows, knowing the name of the intrinsic, you can determine the data type of the input and output data—they aren't necessarily the same. The right-hand column examples are very useful in describing the less obvious functions. For instance, the ABS of a COMPLEX number actually returns a REAL magnitude of the number.

Name	Function definition	In	Out	Example
ABS	Absolute value	I	I	ABS(-5) returns 5
		R	R	ABS(7.7) returns 7.7
		D	D	
		C	R	if CPX = (3.0,4.0) then ABS(CPX) returns 5.0
CHAR	Convert INTEGER to CHARACTER	I	Ch	CHAR(65) returns 'A'. This is provided to circumvent type checking
ICHAR	Convert CHARACTER to INTEGER	Ch	I	ICHAR('Z') returns 90. This is provided to circumvent type checking
LOG	Natural logarithm	R	R	LOG(3.8) returns 1.335. . . .
		D	D	
		C	C	
MOD	Remainder after division	I	I	MOD(14,3) returns 2
		R	R	
		D	D	
SQRT	Square root	R	R	SQRT(81.0) returns 9.0
		D	D	
		C	C	

Pitfalls

There are many ways you can allow "coding" errors to creep into your program when you use subprograms. The best general rule to remember is, *never* allow the I-N rule to be activated. Declare *everything*. And, as long as you are making a declaration, also add a comment line.

Other things have been pointed out to you, but we bring them to your attention once again:

1. Subprogram definitions are not permitted within other definitions— that is, definitions can't be nested. That means that any subprogram can be called from any other subprogram.

2. Main program variables are not available or automatically defined in subprograms as they are in Pascal. You must pass all data into subprograms via the argument list.

3. In FORTRAN, SUBROUTINEs are activated with the CALL verb. Failure to use CALL will lead FORTRAN to think you are executing a FUNCTION and will result in some complicated error messages.

4. You must declare FUNCTION names in the subprogram that invokes that FUNCTION. Failure to do so will lead FORTRAN to think you are referencing an undeclared array.

5. FORTRAN makes no checks to see if the number of arguments in the subprogram declaration and in its usage agree, either in number or in type; that is up to you. This is especially annoying if constants are used when calling the subprogram. If the INTEGER constant 5 is used in the arguments list and the subprogram is declared with a REAL number, the error is not detected.

6. Subprogram names can be no greater than 6 characters, like all other variable names.

Finally, to give you a better understanding of FORTRAN's "weak-typing," the following nonsense program will compile *without* any errors—even though it is full of them, as indicated in the comments. It will not run, of course, and the resulting error messages are very obscure. I hope that this example will serve as a warning to you to be extremely careful when using subprograms and to give you more possibilities of suspicious code to look at when your program refuses to run.

```
        PROGRAM SUBR
C
C SUBROUTINE validity tests
C
        INTEGER J
        REAL A, B
C
C Correct CALL  ...
        CALL MORK (A,B,J)
C
C Not enough arguments  ...
        CALL MORK (A,B)
C
C Arguments out of order  ...
        CALL MORK (J,B,A)
C
C SUBROUTINE used in FUNCTION context  ...
        J = MORK (A,B,J)

        END

        SUBROUTINE MORK(X,Y,I)
C
        INTEGER I
        REAL X, Y

        END
```

Exercises

1. Show how code in the examples (either TRAP or SBTRAP) was derived from the trapezoid rule. Why was a perfectly simple formula coded in such a complicated way? Hint: Start by finding a calculus book. Locate and copy the general form of the trapezoid rule. It is probably expressed as a summation over "n." Expand the summation for a specific value of "n," for instance, $n=5$. Now combine terms, and rearrange them to see if you can get to our solution.

2. Compare the two example Programs (QUAD and SBQUAD). List all the lines that differ. Don't bother with the comment lines.

3. Copy the QUAD example into a file, changing only FUNCTION F as indicated below. Then compile and run it for various other functions over the interval 0 to pi:
   a) Sin (x)
   b) Sin (5x)
   c) e^x
   d) ln (x)
   e) x

   What can be said about the results? Compare the answers to the "right" answer which you can easily compute analytically. Consider the following questions:
   a) Do more iterations *always* result in a more accurate solution? Do the answers converge to the analytical solution for the functions above?
   b) Can you say anything about a minimum number of trapezoids required for 0.001% accuracy for all of these functions? Is the number of trapezoids constant for all those functions? Can you make a hypothesis for monotonic functions?

4. Design/code a function for the "rectangle" rule. Compare the results of the rectangle rule to the results of the trapezoid rule in the previous problem. Consider the following questions:
   a) Compare the magnitude of the error of each method for the various functions. Which rule appears to be more accurate?
   b) Can fewer iterations be used with either method to yield the same accuracy?

5. How would you code FUNCTION F for the function described in Exercise 2, Chapter 2? Do it, and compare the digital solution to the analytical solution.

6. The "error function" has no analytical solution, but often enters into solutions of real-world problems. Furthermore, it isn't part of the FORTRAN library. If you need it, you have to find it in a book of tables. The definition of the "error function" is:

$$\text{erf}(X) = \frac{2}{\sqrt{\pi}} \int_0^X e^{-t^2} dt$$

Modify F (in QUAD) to implement the integral above, and then design/code a main program that will:
a) input a value for X;
b) compute the "erf(X)";
c) display the result, and
d) loop to step (a).
Compare the results of your program to those in a published table. For added interest, write two programs: one with REAL variables and one with DOUBLE PRECISION variables, then compare the results of the two programs to the published tables.

7. One definition of π is expressed below. Using the trapezoid rule, what is the least number of iterations required to accurately express π in 5 digits (i.e., 3.1415)? I was unable to get the sixth digit on my computer (32 bit, single precision). What happens on your machine? If you have the time, continue halving the interval to see how accurate your solution becomes.
   a) What is the smallest number of iterations required for the most accurate solution?
   b) How many digits did your machine compute correctly?
   c) What, do you suppose, is causing this limit?

$$\pi = \int_0^1 \frac{4}{1 + x^2} \, dx$$

8. Repeat Exercise 7 using DOUBLE PRECISION variables instead of REAL variables, answering all the parts to that problem again. Does your machine appear to run slower when operating with DOUBLE PRECISION variables?

9. Design/code a program that accepts a point expressed in either rectangular, cylindrical, or polar coordinates and outputs that point in the other two coordinate systems. This is a good opportunity to test your skills in subdividing your program into functional units—the various conversion equations should appear only once in your final program. Of course you will also need several of the built-in trigonometric routines, and you will need the table in Appendix A to find out how to use them.

10. Find the distance between a line and a point. Write a program that will accept the x_1 and y_1 intercepts of the line and the x_2, y_2

coordinates of the point and then compute and display the distance.

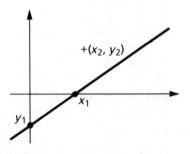

11. Return to Exercise 3 in Chapter 4. There are several parts to it; any one part will do for this problem. For instance, we'll assume you did the 20% part of that problem.

a) Turn the 20% resistor problem into a FUNCTION—call it BEST20—which inputs a resistance value RIN and returns the nearest 20% standard value.

b) Write a SUBROUTINE called TSEC to compute the two resistors of the circuit shown. The SUBROUTINE should have two inputs: Z and K, where $K = E_i / E_o$ and return values for R_1 and R_2. Here is some help:

$$R_1 = Z\,\frac{K - 1}{K + 1} \quad \text{and} \quad R_2 = \frac{2ZK}{K^2 - 1}$$

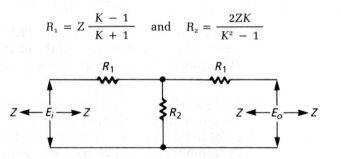

c) Write another SUBROUTINE called PISEC to compute the two resistors for the circuit below. PISEC has the same inputs and outputs as TSEC. The equations are:

$$R_3 = Z\,\frac{K^2 - 1}{2K} \quad \text{and} \quad R_4 = Z\,\frac{K + 1}{K - 1}$$

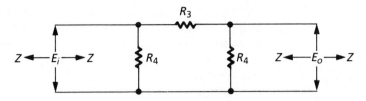

Now, we are ready to put this thing together. The resistor networks shown have the special property that they attenuate while keeping the impedance constant in both directions. However, when translating from theory to a practical circuit, using standard resistors, things go slightly askew. That is what we are going to look at.

Your main program should implement the following algorithm:

a) Prompt/read Z, E_i and E_o.

b) Call TSEC to compute R_1 and R_2.

c) Run R_1 and R_2 through BEST20 to obtain practical values.

d) Compute Z_1, the actual impedance using the practical R_1 and Z_2, the actual impedance using the practical R_2. You can compute these Z's by playing a little algebra on the equations for R_1 and R_2. On the other hand, if you've had Circuits I, you should be able to derive a better equation.

e) Repeat steps (b), (c), and (d) using PISEC instead of TSEC.

f) Print the results in some nice, user-friendly format, then return to step (a).

If you are a hi-fi buff, you might look to see what happens with $Z = 16$ ohms. Telephone freaks would like 600 ohms better. TV folk work with 75 and 300 ohms. But those are just suggestions. Have fun putting this one together!

12. More on numerical analysis. This time I would like to introduce you to the concept of solutions to a polynomial equation of a single variable, like the following:

$$f(X) = C_n X^n + C_{n-1} X^{n-1} + \cdots + C_1 X + C_0$$

which can also be written:

$$f(X) = (\,(\ldots(\,(\,C_n X + C_{n-1})X + \cdots + C_2)X + C_1)X + C_0$$

Remember discussing this form in Chapter 3?

Of course there are "n" solutions to that equation, but I would like to discuss how to get just one real solution. If we can find a "b" for which $f(b) = 0$, then "b" is a solution. Graphically, $f(X)$ crosses the x-axis at "b." We need to discover this crossing. Imagine some value a little less than "b"—call it "b^-"—and another a little larger than "b"—call it "b^+". It is clear that $f(b^-)$ has the opposite sign of $f(b^+)$? If the function "f" is not widely discontinuous, this is true. What we would like to do, then, is to find where $f(X)$ changes sign. Now, let's assume that $f(X)$ crosses the x-axis someplace between "a" and "b." Actually, the tricky part of the algorithm is that you *must* have good values for "a" and "b." If you do, the algorithm will always work—to the limits of your machine. In the following algorithm, $f(a)$ and $f(b)$ are assumed to have opposite signs.

a) Let $A = a$ and $B = b$, assuming $a < b$. Compute $f(A)$ and $f(B)$.
b) Let $C = (A + B)/2$—the middle of the interval—and compute $f(C)$.
c) If the sign of $f(A)$ is the same as $f(C)$, then $A = C$. Otherwise, $B = C$. This step causes the interval to be halved, either from the left or from the right.
d) Test to see if the answer is "good enough." In general we can *never* find an $f(C) = 0$ because REAL numbers on the computer are actually discretes, but we can find something pretty close. Instead we look to see how close A and B are. If $B - A > q$, then go to step (b)—otherwise either A or B is the solution.

The value for "q" depends on your machine and the accuracy that you desire. You can make "q" the smallest REAL number on your machine. And how do you know that? Did you do Exercise 8 in Chapter 3? Of course, you can pick some arbitrary value too, like 10^{-6}.

So, just in case you haven't guessed, you should design/code a FUNCTION—call it REALRT (REAL root)—to implement the algorithm above. Your main program should do the following:

a) Prompt/read "n" and all the "$n+1$" coefficients of the poly-nomial.
b) Prompt/read the end points of the interval, "a" and "b."
c) Call REALRT and then display the answer.
d) Loop to step (b) to change the interval, thus find another root of the same polynomial.

It wouldn't hurt one bit to add some code to assure that $a < b$. You can use MAX and MIN intrinsic functions to guarantee it. And you should probably add code to make sure that $f(a)$ and $f(b)$ really do have different signs, as we have assumed. You can easily test your code by entering a trivial quadratic equation, one with two real roots. You can compute the roots with the quadratic formula to check REALRT.

Note

I don't mean to mislead you into thinking that only polynomials can be solved with this method; any function can be plugged in. But, in the interest of easily entering equations, I choose to use polynomials. The next problem will illustrate this point.

13. Use the half-interval method described in the previous exercise to find one solution to each of the following equations. Since all of these equations have known solutions (although you may have to dig awhile to find them), you should be able to compare your numeric solution to the actuals.
a) $\log_{10}x = 0$
b) $\text{Sin}(x) = 0$
c) $x^3 - x^2 - 9x + 9 = 0$
d) $\text{Sin}(x) / x = 0$
e) $x e^{-x} = 0$

14. In fluid mechanics, pipe-flow problems are solved using the Darcy–Weisbach equation, the continuity equation, and the Moody diagram. A most complex problem arises when only the diameter (D) is the unknown. A textbook solution to this problem is to treat it as a trial and error problem as follows:
a) Knowing L, Q, h_f, g, and v, assume a value for the friction factor (f).
b) Solve the following equation:

$$D = \sqrt[5]{\frac{8LQ^2}{h_f g \pi^2} f}$$

where:

D is the pipe diameter (in feet);
L is the length of the pipe (in feet);
Q is the discharge rate (in cubic feet per second);
h_f is energy loss (in foot-pounds/pound);
g is the gravitional constant, 32.2 feet/second^2;
f is the friction factor.

c) Compute Reynolds' number using the following equation, with the approximate value of D computed from step (b).

$$R = \frac{4Q}{\pi v D}$$

where:

v is kinematic viscosity (in feet/second).

d) Compute the relative roughness, e/D, using published values of e. The variable e is a measure of the size of the roughness projections and has the dimension of length.

e) Enter the Moody diagram using R and e/D to guess at a new value of f.

f) Repeat this algorithm at step (b) until f doesn't change. It is at this point that D, the pipe diameter, has been determined.

You are to design/code a very user friendly program that a mechanical engineer would use to simplify this process. That is, your program should do the following:

1. Prompt/read the necessary constants, converting as necessary. Watch your units!
   i. Discharge rate (in gallons per minute): Q.
   ii. Kinematic viscosity (in feet^2/second): v.
   iii. Pipe length (in feet): L.
   iv. Energy loss (in foot-pounds/pound): h_f.
   v. Roughness (in inches): e.

2. Prompt/read a value for friction factor, f.

3. Make the necessary calculations, and display answers in the proper units for equations (b) with D in inches, (c), and (d).

4. The engineer then uses the Moody diagram (step (e) above) to guess at a new value for f, so your program loops to step 2(b).

Of course, since this problem is in the subprogram chapter, your program should make good use of subprogram structures, and not be a single, monolithic, main program.

6

FORMATTED
INPUT/OUTPUT

The combined concepts of "I/O lists" and FORMAT specifications is probably the most unusual part of FORTRAN. There is no way to cover here all the possibilities of the FORMAT "sub-language," for it is truly a language in itself, so don't be surprised if you see many variations during your career.

Like the DO construct, the FORMAT language has changed little over the generations of FORTRAN. To be sure, some additions have been made but Backus, back in '58, invented the majority of what you will be seeing here. But, before looking at the FORMAT details, there are some details concerning READ/WRITE that need to be understood first.

6.1 THE "READ" AND "WRITE" CONSTRUCTS

The READ and the WRITE constructs, as described in this chapter, are identical. You will see only the "formatted sequential" forms of these statements now. There are other forms; they will be discussed in Chapter 10. For our immediate purposes, the sequential constructs look like this:

READ (unit, format) I/O_list

WRITE (unit, format) {I/O_list}

where the following definitions apply:

- "unit" is either a * (meaning the terminal, as we have been doing) or an integer (meaning a file) as will be discussed in Chapter 10;
- "format" can be either a * (as we have been doing up to this point) or the statement label of a FORMAT statement, as described in Sections 6.3 and 6.4;
- "I/O_list" is an optional list of names of variables, and/or of constants, as described in Section 6.2.

As an introduction, the formatted WRITE statement looks like the following two statements. You're not going to understand what these statements mean yet—be patient.

```
      WRITE (*, 10000) IMAX, ( A(I), I=1, IMAX )
10000 FORMAT ( I5, 8F10.5 )
```

6.2 THE I/O LIST

The "I/O_list" has almost identical forms in both the READ and the WRITE statements. The two differences are pointed out in the list that follows. When the "I/O_list" is longer than a single element, elements are separated with commas, which is just what you've been doing.

The "I/O_list" is permitted to take on one or more of the following forms:

- a simple or subscripted variable;
- an expression, including FUNCTION calls (used in WRITE only);

- a constant, either numeric, logical, or character (used in WRITE only);
- an unsubscripted array name;
- an implied DO.

So far, you have been working with the easiest of the "simple list elements"—that is, the first three items on this list: simple variables, expressions, and constants. Of course, the only places that expressions and constants make sense are in WRITE statements.

UNSUBSCRIPTED ARRAY

When an unsubscripted array name is used in an "I/O_list," the entire array is to be either read in or written out. The format is based on the array name's declaration. Operational details are very different for WRITE than they are for READ, so let's look at a WRITE example first—it's easiest. Suppose there is a FORTRAN program containing the following essential statements:

```
      .
      .
INTEGER ARRAY (1:3,1:2)
DATA ARRAY / -2, -1, 0, 1, 2, 3 /
      .
      .
WRITE (*,*) ARRAY
      .
      .
```

Here is what happens:

- Six numbers will be fetched from the array named ARRAY, formatted as INTEGERs and displayed. In general, the numbers won't all fit on a single line, because FORTRAN will leave enough room for the largest possible INTEGER to be printed.
- The left to right order of the display will be: ARRAY(1,1), ARRAY(2,1), ARRAY(3,1), ARRAY(1,2), ARRAY(2,2), and ARRAY(3,2).

An example of this feature is found with the READ example in the following paragraph.

The READ of an unsubscripted array is much more complicated, because it does so much more for the user/programmer. You may already have discovered many of the features we are about to point out. You will probably find that this is so powerful, in fact, that you may want to use it exclusively for terminal input instead of the formatted READ described in Section 6.4. Here are the essential FORTRAN program statements of a READ example:

```
        .
        .
        .
INTEGER ARRAY (1:3, 1:2)
        .
        .
READ (*,*) ARRAY
        .
        .
        .
```

and here is what happens:

- Six numbers will be fetched from the terminal and stored in the array named ARRAY. The rules for the format of the numbers are pretty loose; almost any number format is accepted as long as at least one blank, or a comma or line separates each number.
- The numbers will be converted to INTEGER format even if you used decimal or scentific notation, but the truncation rule applies if you input data to the right of the decimal point.
- The order of storage into ARRAY will be: ARRAY (1,1), ARRAY(2,1), ARRAY(3,1), ARRAY(1,2), ARRAY(2,2), and ARRAY(3,2). This is the same order that the WRITE statement and the DATA statement use.

To see just how far you can push this feature, I've coded a trivial program for you and then fed it several input numbers. Just read through the examples to get a sense of how "user-friendly" FORTRAN is. Notice that the WRITE examples have been included here too:

```
        PROGRAM I01
C
C Test various array I/O constructions
C
        INTEGER TABLE(1:3, 1:2)

        WRITE (*,*) 'Input 6 values for TABLE'
        READ (*,*) TABLE
        WRITE (*,*) "TABLE is: ", TABLE

        END
```

This is how the program runs on my computer system:

```
$ RUN IO1
Input 6 values for TABLE
1 2 3 4 5 6
TABLE is:        1        2        3        4        5
               6


$ RUN IO1
Input 6 values for TABLE
1,2,3,4,5,6
TABLE is:        1        2        3        4        5
               6


$ RUN IO1
Input 6 values for TABLE
1.1 2.2 3.3 4.4 5.5 6.6
TABLE is:        1        2        3        4        5
               6


$ RUN IO1
Input 6 values for TABLE
1
2
3
4
5
6
TABLE is:        1        2        3        4        5
               6


$ RUN IO1
Input 6 values for TABLE
1 -2 3 -4 5 -6 7
TABLE is:        1       -2        3       -4        5
              -6


$ RUN IO1
Input 6 values for TABLE
1E1 1E2 1E3 1E4 1E5 1E6
TABLE is:       10      100     1000    10000   100000
          1000000
```

(continued)

```
$ RUN IO1
Input 6 values for TABLE
A B C D E F G H

Input conversion error
  unit -4 file SYS$INPUT:.;
```

This last RUN example indicates that nondigits were encountered, when FORTRAN was expecting digits. The error message isn't very illuminating, but you get used to that sort of thing.

IMPLIED DO

Now, let's see just how the "implied DO" is used in an "I/O list" and why. We have seen that the "unsubscripted array name" form was used to transfer an entire array in or out of memory. The "implied DO" form is useful for READing or WRITEing parts of arrays and for incrementing through an array in other than column order. As its name indicates, the "implied DO" is a powerful method of incorporating a DO loop into a READ or WRITE statement. The general form is:

(I/O_list, varib = init, final {, step})

where the following definitions apply:

- "I/O_list" is one or more of the five forms already defined. Notice that this includes "implied DO," so this construct can be nested. Although this looks all very neat, there is one syntactical problem: FORTRAN has no way to distinguish between a nested "implied DO" and an expression enclosed in parentheses. In the latter case, the programmer must flag the expression with a leading + sign. This point is demonstrated in Exercise 4;
- "varib" is an INTEGER or REAL variable;
- "init" is the initial value assigned to "varib";
- "final" is the terminal value of "varib";

- "step" is the value added to "varib" each time through the loop and it is optional. If not specified, it is assumed to be 1. "Step" is not allowed to be zero.

Although similar to the DO construct, there are differences:

- the "body" of the "implied DO" is *before* the DO-like statement;
- the parentheses define the range of the "implied DO," not statement labels;
- there is no statement label in the "implied DO;"
- the body of the "implied DO" is limited to variable and FUNCTION references only; assignment statements are not allowed.

An example is the next order of business. In the code segment that follows, the entire array QTAB is printed, but in row order instead of column order. When "implied DOs" are nested, the punctuation seems especially perplexing—so be alert:

```
      .
      .


REAL QTAB(1:9, 1:14)
      .
      .


WRITE (*,*) (( QTAB(I,J), J=1,14 ), I=1,9 )
      .
      .
```

To see how this example works, consider the following nested DO loop with a WRITE statement, which works similarly. Here is the important difference: The "implied DO" will print as many numbers as possible on a single line, while the nested DO loop example guarantees that only one number per line will be printed. Also, the "implied DO" offers a substantial advantage over the following code, even though both code segments print all of QTAB.

```
      DO 200 I=1, 9
         DO 100 J=1, 14
            WRITE (*,*) QTAB(I,J)
100      CONTINUE
200 CONTINUE
```

6.3 THE "FORMAT" CONSTRUCT FOR "WRITE"

Now we are ready to examine the second big concept. We will begin this subject by describing the commonly used WRITE-FORMAT usage. The READ-FORMAT is introduced in the next section. We have separated the READ- and WRITE-FORMAT discussions because there are so many differences between the two; separate presentation allows us to underscore these differences better. The several "specifications" below are only a small subset of FORTRAN's capabilities. The complete set is contained in an appendix. You may want to refer back to Section 6.1 to review the context of this discussion.

The general form of the FORMAT statement is:

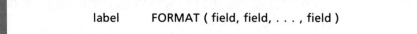

label   FORMAT (field, field, . . . , field)

where the following definitions apply:

- "label" is a positive integer with, at most, five digits;
- "field" is one of the specifications in the description that follows or in Appendix C.

The FORMAT statement is like a DATA statement because it contains a number of constants. But that is where the similarity ends; here are some general points concerning the FORMAT statement:

- The FORMAT statement can be placed anywhere in your program, but a good practice is to put it either with its READ/WRITE statement or in the initialization section of the subprogram or at the very end of your program. Your instructor may guide you on this point.
- The "label" on a FORMAT can't be used in any other context; for instance, it can't be the destination of a GOTO or be used as the end of a DO.
- You should assign FORMAT statement numbers so that they won't be confused with other "labels"; usually I start FORMAT numbers at 10000.

■ The FORMAT statement can be continued like other FORTRAN statements, using column 6. But you should not "continue" a character constant.

A model of the FORMAT mechanism may help you to understand what is happening and give you a better feeling for debugging problems. The FORMAT statement and the READ/WRITE statement are connected via the "label." When the READ/WRITE is executed, the corresponding FORMAT statement is located and the FORMAT interpreter is entered. This interpreter scans the FORMAT statement, from the left, for "fields." Whenever it finds one, it matches it to a variable in the READ/WRITE statement's "I/O_list"; then it performs the conversion indicated by the "field," and finally it transfers the data according to the READ/WRITE function. The interpreter continues scanning the FORMAT statement until one of two things happens:

a) if the interpreter reaches the end of the FORMAT, the FORMAT is repeated—"reviewed"—subject to certain rules, or
b) if the interpreter reaches the end of the "I/O_list," the READ/WRITE statement execution is complete, and execution of the program continues.

You may have noticed that this model permits multiple usage of the FORMAT statement—that is, more than one READ/WRITE can share a common FORMAT statement, but usually this is not done.

Specifically, in the case of the WRITE statement, the "field" describes how the corresponding data are to be converted from internal units (INTEGER, REAL, etc.) to the display format—how many characters are to be generated for the output line, the position of the characters within the specified collection of columns, and various other incidental bits of information, as described here.

Here is another new idea. You are allowed to control the vertical "spacing" of the output lines you are generating. This is called the "carriage control." You can single space or double space your lines, you can force one line to print over the top of another—called "overprint"—and you can cause the paper to be ejected to the perforation. The signal for the carriage control is the first character of the line. Every line has a carriage control on it. Hence, the first character to be *printed* is actually the second one on the line. You must keep this in mind when designing your

WRITE FORMAT. These carriage controls are summarized in the following tabulation:

Plus	(+)	Overprint.	Start the line over again without advancing the carriage.
Blank	( )	Single space.	Start on the next line.
Zero	(0)	Double space.	Skip a line, then begin on following line.
One	(1)	Top of page.	Skip to perforation, then begin at the line following it.
Others		Same as a blank.	

Of course, you must consider the destination device when choosing carriage controls. The "top of page" doesn't make a whole lot of sense on a CRT terminal, and "overprinting" on a CRT simply doesn't work. These concepts are illustrated in the first example and in the exercises.

We are now ready to look at the details of the "fields." The general form of a FORMAT "field" (with several exceptions) is:

$$r f w . d$$

where the following definitions apply:

- "r" is a positive integer, the "repeat" count. It is optional; if missing, it is assumed to be unity;
- "f" is the "field" code, described below;
- "w" is a positive integer, the field "width";
- "." is required punctuation for some "fields."
- "d" is a non-negative integer which stands for number of "digits" to the right of the decimal.

In the examples below, the symbol ƀ means "blank space."

A FIELD   {r}A{w}

The A field descriptor transfers *characters* to the print line without conversion. If the field width "w" is wider than the definition of the corresponding variable, the leading blanks are supplied. If "w" is too small, only the leftmost "*w*" characters are transferred. If "*w*" is not present, the defined length of the corresponding variable or constant is used. Therefore, there is no way to make an error using the A field descriptor without the "*w*."

Field	Memory value	Output	Comments
A6	Ohms	ƀƀ Ohms	Leading blanks are added
A5	Volts	Volts	Exactly right
A4	Amperes	Ampe	Not enough room, the string is chopped off on the right.

F FIELD   {r}Fw.d

The F field descriptor causes rounding, conversion, and transfer of REAL and COMPLEX variables to the output line in decimal notation. The total field width is defined by "*w*" and the number of digits to the right of the decimal point is defined by "*d*," therefore "*w*" must be greater than or equal to "*d*" + 3. The displayed number is right-justified in the "*w*" positions, with leading blanks if necessary.

Field	Memory value	Output	Comments
F8.3	2.4567654	ƀƀƀ2.457	Value is rounded, leading spaces.
F7.2	−987.1234567	−987.12	Exactly enough room.
F7.3	−987.1234567	*******	Error; not enough room.
F8.2	−.2	ƀƀƀ−0.20	Notice zero left of decimal.

I FIELD   {r}Iw{.m}

The I field descriptor converts and transfers INTEGER variables to the output line. The digits will be right-justified in the "*w*" positions. If the number doesn't fill the "*w*" positions, leading blanks are transferred. The optional "*m*" field causes leading zeros (not blanks) in the rightmost "*m*" positions; ("*w*" − "*m*") are blank filled.

Field	Memory value	Output	Comments
I3	456	456	
I2	234	**	Error; not enough room
I4	−123	−123	
I5	987	ƀƀ987	Two leading blanks are inserted
I4.2	6	ƀƀ06	A leading zero and two leading blanks are inserted
I3.3	−65	−65	The sign takes the place of the leading zero

T FIELD   T*n*

The T field descriptor defines a position, "n," to place the next output character. No actual output is generated with this descriptor. Notice that the carriage control character is considered position 1, so the first printable character position is 2. This descriptor, in effect, allows a "position pointer" to move both right and left on the output line.

X FIELD   *n*X

The X field descriptor moves the "position pointer" "n" columns to the right. There is no output for this descriptor.

SLASH FIELD   /

The slash terminates output of the current line and initiates a new one. Sometimes it is also used as a sort of carriage control, since a double slash will cause one line to be skipped. If a slash is used, the comma separator is optional, but you will notice in the following examples that commas make the FORMAT more "readable." The slash isn't required as the final character of a FORMAT—the right parenthesis will terminate the line—but it is permitted.

CONSTANTS

Displaying constants can be done two ways: either with or without a field descriptor. Therefore, constants can be included either in the WRITE statement or in the FORMAT statement. The two methods are illustrated here.

To output a constant number:

```
      WRITE (*,11000) 84
11000 FORMAT ( I3 )
```

and

```
      WRITE (*,11001)
11001 FORMAT ( ' 84' )
```

will produce the same display. Notice that the character string had to be enclosed in apostrophes in the 11001 FORMAT statement, but not when the FORMAT interpreter did the "I" conversion in the WRITE (*, 11000).

The same principle applies to constant character strings. Here are two examples:

```
      WRITE (*,12120) ' Moment of Inertia'
12120 FORMAT ( A )
```

and

```
      WRITE (*,12301)
12301 FORMAT ( ' Moment of Inertia' )
```

In the first case the A field was used and allowed "*w*" to be defined from the constant itself. In the second case, the constant string was stored in the FORMAT statement and no FORMAT field was necessary. Incidentally, it is the second form that cannot be continued from one line to another. In all four examples the first character—a blank—was used as carriage control.

One last topic: two forms of repetition within the FORMAT statement. The first, repeat counts, in the form:

```
20200 FORMAT ( F13.5, F13.5, I6, I6, I6 )
```

could be written as:

```
20200 FORMAT ( 2F13.5, 3I6 )
```

which, clearly, is easier to read and write—a real improvement. The second form of repetition is called "repeat groups." In this form, you are allowed to group fields with parentheses and add a repeat count to the combined fields. For example, to process six variables:

```
30010 FORMAT ( '0', F13.6, 2X, I4, F13.6, 2X, I4,
     +            F13.6, 2X, I4 )
```

can be written as:

```
30010 FORMAT ( '0', 3(F13.6, 2X, I4) )
```

Another feature in the FORMAT interpreter involves implied repetition: If the interpreter gets to the end of the FORMAT before getting to the end of the I/O list, the FORMAT is repeated. This is called FORMAT "reversion." The rules in this case are:

1. A new line is started, so a carriage control is required.
2. The FORMAT is restarted at the left parenthesis corresponding to the next-to-the-last right parenthesis and includes the repeat count, if any.

This is a complex subject, and an example will be helpful. You are shown above, in 30010, how to process six variables without reversion. But, suppose there were actually 10 variables in a WRITE I/O list instead of six; the first six would appear on the first line, and the remaining four on the second line. If the first version of 30010 were used to WRITE the 10 variables, the '0' would be used for the carriage control. But if the second version of 30010 were used, the second line's carriage control is unknown because of rule 2. The whole FORMAT is not repeated, only up to the inner left parenthesis. Let's try it. Look at the following program, simplified to make the point:

```
        PROGRAM I03
C
C A program to illustrate how 'FORMAT reversion' works
C
C Display 10 values with simple reversion
C
        WRITE (*,30010) 1., 1, 2., 2, 3., 3, 4., 4, 5., 5
30010   FORMAT ( '0', 3(F13.5, 2X, I4) )
C
C Display 10 values by repeating the whole FORMAT
C
        WRITE (*,10001) 1., 1, 2., 2, 3., 3, 4., 4, 5., 5
10001   FORMAT ( ('0', 3(F13.5, 2X, I4) ) )

        END
```

```
$ RUN IO3
       1.00000      1      2.00000      2      3.00000      3
       4.00000      4      5.00000      5

       1.00000      1      2.00000      2      3.00000      3

       4.00000      4      5.00000      5
```

You see 30010 again, only this time it is paired to an I/O list with 10 elements in it. Hence the group repeat must be revisited. This means that the carriage control character is missed on the second line; therefore the second line is shifted left one character. This was corrected in 10001 by adding a second set of parentheses (reversion goes back only one set of parentheses) to force reversion back to the beginning of the FORMAT, so that the output lines up in columns. Why don't the two rows "line up" in the first case? Because no carriage control is specified for the second line. The first blank of the number field is used instead; therefore it is not printed.

There are other solutions to this problem and many other tricks that can be invented once you understand "reversion" logic.

6.4 THE "FORMAT" CONSTRUCT FOR "READ"

As mentioned, READ/FORMAT isn't very convenient for screen input, but rather for file input, which is covered in Chapter 10. Interactive, terminal input should always use READ (*,*) . . . code, because it is so much more user-friendly. The READ/FORMAT section is included here for the sake of completeness only; once you've mastered WRITE/FORMAT, this topic will present no problems.

This section describes the most commonly used READ/FORMAT "specifications." Some of them, like F and I, work differently, so be alert. As in the WRITE/FORMAT section, the several "specifications" have been arranged alphabetically for easier reference.

Here, once again, is the general form of the FORMAT statement:

> label   FORMAT (field, field, . . . , field)

where the following definitions apply:

- "label" is a positive integer with no more than five digits;
- "field" is one of the specifications in the description that follows or appears in Appendix C.

The major difference in the operation of the READ/FORMAT over the WRITE/FORMAT is that data are being transferred *from* character form *to* internal form. We still talk about "*w*," the width of the character field, but "*d*," the number of characters right of the decimal, has a new meaning, which will be described shortly. Usually, when looking at a FORMAT statement, it is difficult to tell if it belongs to a WRITE or a READ statement.

A FIELD $\{r\}A\{w\}$

The A field descriptor transfers "*w*" characters into a CHARACTER variable. Characters are transferred from the left—that is, the first character to be transferred is the leftmost. Then, "*w*" is used to establish how many characters to move to memory. If "*w*" is omitted, the declaration of the variable is used to find "*w*."

Field	Input	Memory value	Comments
A5	Windows	Windo	Input chopped off
A7	Windows	Windows	All characters transferred

F FIELD $\{r\}Fw.d$

The F field descriptor converts "*w*" characters to REAL format and stores the resulting value in memory. Any input decimal notation is acceptable, but if the decimal is missing, the rightmost "*d*" digits are considered to be right of the decimal point (see the first two examples below).

Field	Input	Memory value	Comments
F7.4	1234567	123.4567	Decimal is assumed
F7.4	12345ƀƀ	123.45	Trailing zeros are assumed

Field	Input	Memory value	Comments
F7.4	ƀ−12.3ƀ	−12.3	Decimal is used
F7.4	−6.2E3ƀ	−6200.0	E notation accepted
F7.4	−6.23D5	−623000.0	D notation accepted

I FIELD   {r}Iw{.m}

The I field descriptor converts "*w*" characters to INTEGER format and stores the resulting value in memory. The character string must be in IN-TEGER FORM; it cannot contain a decimal point or an exponent field. The optional ".*m*" part is ignored if present. It is accepted by FORTRAN so the FORMAT can also be used in a WRITE statement.

Field	Input	Memory value	Comments
I4	−12ƀ	−12	Trailing blanks are ignored
I5	ƀƀ97ƀ	97	The number can be anywhere in the "*w*" character field

T FIELD   Tn

The T field descriptor defines a position, "*n*," for the location of the next input character. No variable in the I/O list is associated with this field. This descriptor allows the programmer to skip around in the input line or even to reprocess a field rather than to process it in strictly left-to-right order. But notice that, unlike output, the input line has actual data in column 1—the leftmost column.

X FIELD   nX

The X field descriptor moves the position pointer right "*n*" columns. This is, in effect, a "relative tabulator," whereas the T descriptor is an "absolute tabulator." The X descriptor allows the programmer to skip fields when reading through the input line.

SLASH FIELD   /

The slash terminates input of the current line and reads a new one. Its primary purpose is to process more than one line with a single READ statement. If a slash is used, the comma separator is optional, but I think that commas make the FORMAT more "readable." The slash isn't required as the first character of a FORMAT; the left parenthesis will cause a new line to be read.

Repeating fields and groups in the READ/FORMAT statement is analogous to repeating under WRITE/FORMAT, and the same rules apply.

Put It Together

There are two examples in this section. The first deals with character manipulations and the second with REAL number variations.

The following program will display the Bemidji State University logo in a 19 × 32 character area, one WRITE-FORMAT for each of the 19 lines. This example is much the same as Exercise 7. However, for the sake of illustration, the various lines have been produced using different FORMAT field techniques, which means that some lines have very messy FORMAT statements, and others look much neater. I am of the opinion that detailed algorithms aren't required for programs like these, but I have added comments whenever they seemed to clarify the point I was making in the particular FORMAT statement.

About the sixth line in the program you will see a surprise: The OPEN has not been discussed yet, and won't be until Chapter 10. Here is the problem: I want to illustrate "overprinting," but my terminal is a CRT, and "overprinting" is, of course, impossible. So, I have to use the OPEN to direct the output to the system's printer. You can accept this OPEN statement on blind faith, or perhaps your instructor will help you find out how to use the OPEN on your system to do the same thing—every system is different.

If you are working with a printing terminal, you have no problem. The OPEN can be removed, and you can change all the WRITE (6,...) to WRITE (*,...).

```
        PROGRAM LOGO

C       A program to print the BSU logo of pines and star.
C       This program partially overprints the logo.

C       Attach unit 6 to the printer - this is MS-DOS specific
        OPEN (6, FILE = 'PRN:')
C
C The first line is double spaced to set it off from other printouts
        WRITE (6,10001)
10001   FORMAT ( '0', T8, 20('X') /
     +  '+', T8, 20('M') / '+', T8, 20('0') )
C
C Put the constants in the WRITE statement instead of the FORMAT
        WRITE (6,10002) 'X', '|', 'X', 'M', 'M', '0', '0'
10002   FORMAT ( T7, A, T17, A, T28, A /
     +  '+', T7, A, T28, A / '+', T7, A, T28, A )
        WRITE (6,10003)
10003   FORMAT ( T6, 'X', T11, 'X', T16, '\|/', T23, 'X', T29, 'X' /
     +  '+', T6, 'M', T11, 'M', T23, 'M', T29, 'M', /
     +  '+', T6, '0', T11, '0', T23, '0', T29, '0' )
        WRITE (6,10004)
10004   FORMAT ( T5, 'X', T11, 'X', T15, '--X--', T23, 'X',
     +  T30, 'X' /
     +  '+', T5, 'M', T11, 'M', T17, 'M', T23, 'M', T30, 'M' /
     +  '+', T5, '0', T11, '0', T17, '0', T23, '0', T30, '0' )
        WRITE (6,10005)
C
C Use blanks instead of tabs for the first X on the line
10005   FORMAT ( '    X', T11, 'X', T16, '\|/', T22, 'XX', T26, 'X',
     +  T30, 'X' /
     +  '+    M', T11, 'M', T22, 'MM', T26, 'M', T30, 'M' /
     +  '+    0', T11, '0', T22, '00', T26, '0', T30, '0' )
        WRITE (6,10006)
10006   FORMAT ( T5, 'X', T11, 'XX', T17, '|', T22, 'XX', T26, 'X',
     +  T30, 'X' /
     +  '+', T5, 'M', T11, 'MM', T17, '|', T22, 'MM', T26,
     +  'M', T30, 'M' /
     +  '+', T5, '0', T11, '00', T17, '|', T22,
     +  '00', T26, '0', T30, '0' )
        WRITE (6,10007)
```

(continued)

```
10007    FORMAT ( T5, 'X', T11, 'XX', T21, 'XXX XX', T30, 'X' /
     +   '+', T5, 'M', T11, 'MM', T21, 'MMM MM', T30, 'M' /
     +   '+', T5, '0', T11, '00', T21, '000 00', T30, '0' )
         WRITE (6,10008)
10008    FORMAT ( T5, 'X', T11, 'XXX', T21, 'XXX XX', T30, 'X' /
     +   '+', T5, 'M', T11, 'MMM', T21, 'MMM MM', T30, 'M' )
         WRITE (6,10009)
10009    FORMAT ( T5, 'X', T11, 'XXX', T21, 7('X'), T30, 'X' /
     +  '+', T5, 'M', T11, 'MMM', T21, 7('M'), T30, 'M' )
         WRITE (6,10010)
10010    FORMAT ( T5, 'X', T10, 5('X'), T21, 7('X'), T30, 'X' /
     +   '+', T5, 'M', T10, 5('M'), T21, 7('M'), T30, 'M' )
         WRITE (6,10010)
         WRITE (6,10012)
10012    FORMAT ( T5, 'X', T10, 5('X'), T20, 9('X'), T30, 'X' /
     +   '+', T5, 'M', T10, 5('M'), T20, 9('M'), T30, 'M' )
         WRITE (6,10013) 'X', 'XXXXXXX', 'XXXXXXXXX', 'X'
10013    FORMAT ( T5, A, T9, A, T20, A, T30, A )
         WRITE (6,10013) 'X', 'XXXXXXX', 'XXXXXXXXX', 'X'
         WRITE (6,10015)
10015    FORMAT ( T5, 'X', T12, 'X', T23, 'X', T26, 'X', T30, 'X' )
         WRITE (6,10016)
10016    FORMAT ( T6, 'X', T12, 'X', T23, 'X', T26, 'X', T29, 'X' )
         WRITE (6,10017)
C
C The next line is formed right to left-just to be different
10017    FORMAT ( T28, 'X', T26, 'X', T23, 'X', T12, 'X', T7, 'X' )
         WRITE (6,10018)
10018    FORMAT ( T8, 'X   X', T23, 'X  XX' )
         WRITE (6,10019)
10019    FORMAT ( T9, 'XXXX', T23, 'XXXX', /, '0' )

         END
```

When LOGO is executed, the pattern shown at the top of the opposite page is displayed.

This second example deals entirely with REAL numbers. COMPLEX numbers are covered in Chapter 8 and DOUBLE PRECISION is left for an exercise at the end of this chapter. The next program merely prints the several REAL numbers in different formats using F, E, G, and P field descriptors. You will have to reference Appendix C to read

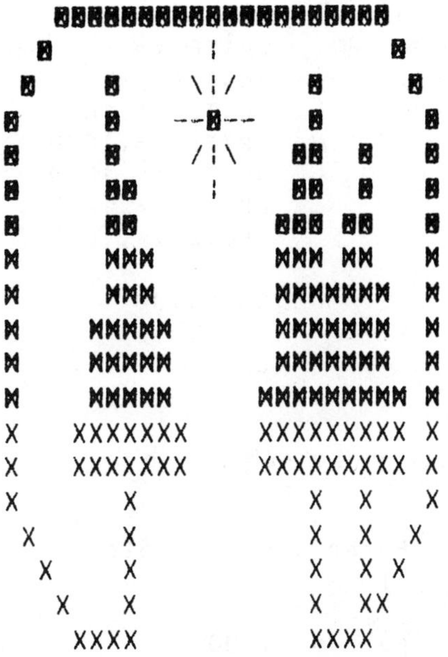

the description of the E, G, and P descriptors. Each output line presents the same number in a different format. As in the previous example, I've tried to show you the variations possible, rather than the "best" method. When the error flag, "*****", is printed in the place of a number, the interpreter is signifying that the number was too big to fit in the specified field.

```
      PROGRAM IO2
C
C Program to read a REAL number, then display it in
C different formats.
      REAL X

10000 FORMAT ( '0F6.2 = ', F6.2 )
10001 FORMAT ( ' F6.2 with sign = ', SP, F6.2 )
10002 FORMAT ( ' E15.7 = ', E15.7 )
10003 FORMAT ( ' G15.7 = ', G15.7 )
10004 FORMAT ( ' 1PE15.7 = ', 1PE15.7 )
```

(continued)

```
100      CONTINUE
         WRITE (*,*) ' Enter a number please'
         READ (*,*) X
         WRITE (*,10000) X
         WRITE (*,10001) X
         WRITE (*,10002) X
         WRITE (*,10003) X
         WRITE (*,10004) X

         GOTO 100

         END
```

Examples of output with this program follow:

```
Enter a number please
-1

F6.2 = -1.00
F6.2 with sign = -1.00
E15.7 =    -0.1000000E+01
G15.7 =      -1.000000
1PE15.7 =    -1.0000000E+00

Enter a number please
1

F6.2 =    1.00
F6.2 with sign =    +1.00
E15.7 =    0.1000000E+01
G15.7 =      1.000000
1PE15.7 =    1.0000000E+00

Enter a number please
123

F6.2 = 123.00
F6.2 with sign = ******
E15.7 =    0.1230000E+03
G15.7 =      123.0000
1PE15.7 =    1.2300000E+02
```

```
Enter a number please
0.12345

F6.2 =    0.12
F6.2 with sign =  +0.12
E15.7 =    0.1234500E+00
G15.7 =    0.1234500
1PE15.7 =    1.2345000E-01

Enter a number please
0.000056

F6.2 =    0.00
F6.2 with sign =  +0.00
E15.7 =    0.5600000E-04
G15.7 =    0.5600000E-04
1PE15.7 =    5.6000001E-05

Enter a number please
-1E30

F6.2 = ******
F6.2 with sign = ******
E15.7 =   -0.1000000E+31
G15.7 =   -0.1000000E+31
1PE15.7 =   -1.0000000E+30

Enter a number please
1E-30

F6.2 =    0.00
F6.2 with sign =  +0.00
E15.7 =    0.1000000E-29
G15.7 =    0.1000000E-29
1PE15.7 =    1.0000000E-30

Enter a number please
0.123456789
```

(continued)

```
F6.2 =     0.12
F6.2 with sign =   +0.12
E15.7 =    0.1234568E+00
G15.7 =    0.1234568
1PE15.7 =    1.2345679E-01
```

Pitfalls

The FORMAT statement gives you many new and exciting ways of making mistakes that FORTRAN cannot detect during compilation, thus giving you an opportunity to exercise your debugging techniques. There is no way to anticipate the troubles you will experience, but here are some of the more common problems:

1. The number of variables in the I/O list doesn't match the number of fields in the FORMAT statement. The problem is that, because of reversion, the number of terms in the I/O list won't necessarily be the same as the number of field specifications in the FORMAT statement; but it is very desirable to make it your goal to have the same number of terms until you understand reversion pretty well.

2. A second common problem is that the field descriptor doesn't agree with the variable's definition—for instance, printing an INTEGER variable using an F field. The problem here is that perfectly valid results can be observed (a 0.0 for example) even though it is wrong.

3. When WRITEing, remember that the first character on the line is always considered a carriage control. If you're not careful, a minus sign can easily get lost this way—likewise, a leading digit. The leading zero on REAL numbers causes double spacing. The leading 1 causes page ejects and your output has only one line per page. This can be very frustrating, because usually it causes the printer to jam over and over again.

4. You may have noticed that FORMATs for READing don't need to allow for spaces between fields. But you must be careful to allow for spaces when designing a FORMAT for WRITEing; pushing numbers and/or letters together doesn't look very nice.

5. Occasionally, you spend lots of time designing your FORMAT statement but forget to include the READ or WRITE statement that causes the actual data transfer. Remember that the FORMAT statement is nothing more than arguments for the READ or WRITE system subroutine; the FORMAT statement cannot stand by itself.

Exercises

1. The use of READ (*,*) shown in this chapter varies greatly, depending on the system you are using. For instance, the examples in Section 6.2 and the "Put It Together" section were run on VAX-FORTRAN®, but when I tried the same thing on MS-FORTRAN®, some of the inputs weren't accepted. Generally, specific manufacturers add a few bells and whistles that are either beyond the FORTRAN-77 standard or that are not covered by it.

This is an open question. Try to discover some other "interesting" things about READ (*,*) on your system. There are no "wrong" answers to this question and it is difficult to know when you are done. Be prepared to share your discoveries with the others in your class.

2. Complete the following table. The body of the table is the number of BTUs lost through infiltration (HL_f) in a house at 1 ACH (air change per hour) in a 24-hour period. The table below assumes that all houses have eight-foot ceilings, so the volume is easily computed. The formula for the body of this table is:

$$HL_f = W * s * t * V * T$$

where:

> HL_f is the hourly heat loss through infiltration (BTU hrs);
> W is the weight in pounds of a cubic foot of 70 degree air (0.077 lb./cu. ft.);
> s is the specific heat of a cubic foot of air (0.24 BTU/(lb. deg));
> t is the difference in temperature between the inside and the outside (°F);
> V is the volume of air (cu. ft.);
> T is the time (hrs.).

Sq Ft	Outside temperature					
	− 20	− 10	0	10	20	30
1000						
1200		340623				170312
1400						
1600						
1800						
2000						
2200						
2400						
2600						
2800						
3000						

You are to design/code a program to compute the values for the body of this table, printing it out in a nice neat way. Approach the assignment in this way: make all necessary computations and store the answers in an array, then print the array without using an implied DO.

3. Design/code a program to compute and store values for the table described in Exercise 2, then print the table using implied DO loops wherever possible.

4. Design/code a program to compute values for the table in Exercise 2, printing as the computations are made—that is, without storing the data in an intermediate array.

5. This is primarily a research problem associated with Exercise 2. The first part of this problem is to verify the constants given above: the weight of a cubic foot of air and its specific heat. The second part of this problem is to determine how the humidity of the air inside the house affects the table. It seems that the weight and specific heat are both functions of humidity. That is what you are to find out. Present your conclusions in a tabular form of your own design, just as in the previous problem. For instance, you may want to produce several "Exercise 2-like" tables, one for 30% relative humidity, one for 40%, etc.

6. This is a typical "professional" problem, because its decription is quite straightforward. Yet, the solution will require some digging on your part. I spent an hour or so looking for the necessary tables in my library, without success. I'm sure you can do better in your campus library. Here is the problem:

Your manager has asked you to produce a set of stall speed (V_s) tables for a private aircraft with the following characteristics:

- weight (W): 10,000 pounds;
- wing area (A): 350 square feet;
- wing lift coefficient (C_{lm}): 3.33.

What he would like is four tables, one for every inch of barometric pressure, from 29 to 32. Each table body should contain the stall speed (to the nearest whole MPH). The rows of the table should show wet-bulb temperatures, and the columns should show dry-bulb temperatures. The formula is really trivial:

$$V_s = \text{SQRT} \left(\left(2 * W \right) / \left(r * A * C_{lm} \right) \right)$$

where the only undefined symbol is "r"—the density of air in pounds per cubic foot. The hardest part of this problem is to find the temperature–relative humidity relationship for air. When you find the right tables, make sure you get the right units!

Of course, your output tables should look professional: columnar, easy-to-read, and so forth. Neatness counts! Add lines to the table, using the minus (−) and bar (¦) characters if it looks nicer; line up the decimal points and double space if that makes it easier to read. Remember, your raise might depend on how good these tables look—and on their accuracy, too, of course.

7. Design/code a program, using many, many WRITE and FORMAT statements that will display your school logo on your CRT terminal, completely filling the screen except for the last line. If you don't use a CRT terminal, refer to the next problem instead. I suggest you design the logo display using a piece of quad-ruled paper first. Outline a workspace the size of your screen (probably 20 × 80), then decide how you're going to fill it, which characters would look best, and so forth. Don't overuse blanks in your FORMAT statements, because this is a good exercise for the T or X descriptor. You should use as much of the screen as possible to make your logo look as nice as possible.

8. Repeat Exercise 7 except that your program should print the logo on hard copy. You will have to discover how (or if) the OPEN on your brand of FORTRAN works—or you may not need it. Your design should be shaded using overprinting, not be a mere copy of the terminal program in Exercise 7. Your printed logo should look much better than a terminal version.

9. This problem may be familiar; it is much like Exercise 7 in Chapter 3. We would like you to investigate integers and their inverses—in REAL and DOUBLE PRECISION. You are to fill in the following table. Use the exact format in your E and D FORMAT descriptions.

	Real		Double precision	
N	1/N	Error	1/N	Error
2	0.500000E+00	0.000000E+00	0.5000000000000000D+00	0.000000D+00
3				
.				
.				
.				
99				

where the error column is computed as follows:

$$error = ((1.0 / N) * N) - 1.0$$

There is one "trick" you must know: to force DOUBLE PRECISION computations, you must use 1.0D0 (1.0 in DOUBLE PRECISION talk) instead of simply 1.0 (which is a single precision quantity) in the inverse computation.

There are three ways to code this problem:

a) Make all computations, store the answers in REAL and DOUBLE PRECISION matrices, then print the table.

b) Using a DO loop, compute everything needed for one line, then print that line.

c) Using an implied DO, compute and print a line using a single WRITE (*,*) statement—no DO loops. This method could be done with or without FORMAT reversion. This method, although somewhat complex, is certainly shorter than either suggestion (a) or (b).

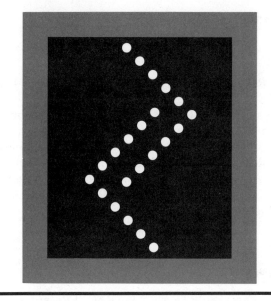

7

"CHARACTER" OPERATIONS

Now that FORMATs have been introduced and you have mastered at least parts of it, we can return to fairly familiar territory: working with CHARACTERs. You have probably already worked with CHARs in Pascal. Now you will see how it is done in FORTRAN.

We have made a special case of this topic because very little of the subject falls into the general scheme of FORTRAN. In fact, this is a "recent" add-on to the language. The first FORTRAN had nothing specific for manipulation of CHARACTERs; a programmer had to work with INTEGER variables instead. Since FORTRAN was developed for number crunching, nobody at that time could foresee CHARACTER pushing. By the time FORTRAN 77 was developed, however, operations on strings of

143

CHARACTERs were commonplace and, of course, belonged in FOR-TRAN too.

We have subdivided CHARACTERs into two sections: working with single CHARACTER variables, and working with strings of CHARAC-TERs. You will learn to declare CHARACTER variables, to manipulate CHARACTER expressions, and to extend your knowledge of input and output of CHARACTERs. Exercises in this chapter deal mainly with display and hardcopy plotting, because it is a very common method of presenting engineering data.

7.1 "CHARACTER" MANIPULATIONS

In their basic form, FORTRAN CHARACTERs are exactly the same as Pascal CHARs. Introducing the topic in this way permits CHARACTERs to be approached universally, across all computer implementations of FORTRAN. But, here is the problem: When FORTRAN is installed on a microcomputer, usually, because of the size of the compiler, something has to be left out. For instance, in MS-FORTRAN®, CHARACTER operations—which we will look at later—were left out entirely, which means that CHARACTER strings are not very easy to use. The point is that, in its basic operations, FORTRAN and Pascal view CHARACTERs identically. Even so, some manufacturers leave out some of these operations.

Let's start with an example. Declaring a simple variable is a trivial case. To set up an array of CHARACTERs is a little more interesting, for instance, in Pascal:

```
name: ARRAY [...] OF CHAR;
```

becomes, in FORTRAN:

```
CHARACTER name (...)
```

To illustrate this method, even though it should be very familiar, suppose you wanted to set up a two-dimensional array the size of your CRT screen. The FORTRAN declaration would be:

```
CHARACTER SCREEN (1:80, 1:24)
```

To put an 'X' in column 40, row 17, the assignment statement would be:

```
SCREEN (40,17) = 'X'
```

And, to display this screen image, the simple statement below will "almost" work. You can examine this in detail in Exercise 9.

```
WRITE (*,*) SCREEN
```

What could be easier! But notice that I've reversed the normal definition of "row" and "column" to force this simplification.

7.2 STRING MANIPULATIONS

If, on the other hand, you are fond of PACKED ARRAY [. . .] OF CHAR, this section is just for you. However, we need to caution you again that these features are included only in the "full" FORTRAN 77 and not in the "subset" of FORTRAN 77 found on most micros.

The general form of a CHARACTER declaration follows. Notice that this is additional detail missing from the general form of declarations shown in Chapter 4:

CHARACTER{*n} var_list

where the following definitions apply:

- "n" is the length of the CHARACTER string. If not specified, "n" is assumed to be 1, exactly like Pascal's CHAR. There is a special case when "n" is "(*)", that is:

```
CHARACTER*(*) ...
```

This notation is allowed only for subprogram argument definitions and means that the subprogram is to use the length of the string as defined in the calling program. This is illustrated in the "Put It Together" section.
- "var_list" is one or more variable names, separated by commas.

So far, this is rather simple. An example follows. We have already seen what happens when the "*n" is not present, so let's look at the other case. The following Pascal declaration:

```
name: PACKED ARRAY [1..10] OF CHAR;
```

in FORTRAN would become:

```
CHARACTER*10 name
```

When declaring higher dimension arrays, the regularity found in Pascal is lost in FORTRAN. Potentially, this can be a cause for confusion to the Pascal programmer. For example, look at the screen image example again. When using strings instead of arrays of CHARACTERS, the declaration using Pascal is:

```
SCREEN: PACKED ARRAY [1..80, 1..24] OF CHAR;
```

But in FORTRAN it would be:

```
CHARACTER*80 SCREEN(1:24)
```

The way you can think about this is that FORTRAN allows you to declare "strings" of character data and arrays of "strings." For example, the code above declares an array containing 24 strings, each 80 characters long.

It is at this point that we deviate from Pascal. Starting with an assignment statement, for example, in Pascal:

```
SCREEN [40,17] := 'X';
```

becomes, in FORTRAN:

```
SCREEN (17) (40:40) = 'X'
```

This code is storing data in the 17th string and the 40th character, which may appear "backwards" from your Pascal training. Let's look at the generalization of a substring reference:

$$\text{varib(expr}_1 : \text{expr}_2)$$

where the following definitions apply:

- "varib" is the CHARACTER variable name, including subscripts if applicable;
- "expr_1" is an integer expression that specifies the leftmost CHARACTER position of the substring. Notice that the first character in the string is called number one.
- "expr_2" is an integer expression that specifies the rightmost CHARACTER position of the substring.

Notice that "expr$_1$" and "expr$_2$" specify an inclusive range, which means that if you are referencing one CHARACTER, "expr$_1$" and "expr$_2$" are identical. And if you are talking about "n" CHARACTERs, then n = "expr$_2$" − "expr$_1$" + 1. Also notice that "expr$_1$" is always less than or equal to "expr$_2$."

Brief examples are shown below, and more complete examples are shown in Section 7.5. Suppose there is a CHARACTER*12 variable in your program, and it is initialized in the following statement:

```
TITLE = 'WINGb SPAN'
```

Since only nine characters are defined by the constant string, the right-most three in TITLE are set to blanks. In the following table, operations and results are shown:

Operation	Results
TITLE(1:4)	'WING'
TITLE(5:10)	'bSPANb'
TITLE	'WINGbSPANbbb'
TITLE(10:12)	'bbb'

Logically, if you can tear strings apart, you should be able to put them together too. There is one operation to do that job. It is called "concatenation" and is written as a double slash: "//". None of the arithmetic operations, e.g. "+", "−", "*", "**" or "/", will work with CHARACTER variables. I have extended the previous example below:

Operation	Results
TITLE(1:4) // '−' // TITLE(6:9)	'WING-SPAN'
TITLE(6:9) // 'bofbtheb' // TITLE(1:4)	'SPANbofbthebWING'

As you might expect, you can compare strings and substrings too. You are allowed to use the familiar .EQ. to test for equality and .NE. to test for inequality of two strings. Normally, testing strings of CHARACTERs makes sense only when the strings are of equal length, but FORTRAN is pretty forgiving. If you test strings of unequal length, the shorter one has blanks concatenated on the right end, temporarily, before the comparison is made.

You *may* use the other relationals too—that is .LT., .LE., .GE., and .GT.—but you have to be aware that not all manufacturers' collating sequences are the same. For example, CDC® defines letters first, then numbers; DEC® defines numbers before letters. Therefore, if you use these relationals, you have to understand the meaning of the results on your machine. In order to avoid this problem, intrinsics have been defined for this purpose.

Several built-in functions are available in standard FORTRAN 77 for working with CHARACTERs. Not all of them are defined on micro FORTRANs however. You will have to check your manual for details.

- CHAR(int) is just like Pascal's CHR; it makes a CHARACTER out of an integer.
- ICHAR(char) is just like Pascal's ORD; it makes an integer out of a CHARACTER.
- INDEX($string_1$, $string_2$) searches "$string_1$" for the leftmost occurrence of "$string_2$" and returns an integer which is the position of "$string_2$". If the search is unsuccessful, zero is returned; the first character of a string must be position one.
- LEN(string) returns the count of the number of CHARACTERs defined in the declaration of "string." "String" is allowed to be an expression, for instance involving the concatenation operator.
- Lxy($string_1$, $string_2$) compares two strings and returns a logical .TRUE. or .FALSE.. There are really four functions; "xy" can be either LT, LE, GE, or GT; the periods aren't used. This is the correct way of doing the .LT., .LE., .GE., and .GT. relationals.

7.3 "CHARACTERs" IN "FORMATs"

When dealing with the A descriptor in FORMAT statements, there are some rules that are specific to that case only and that were not discussed in the previous chapter. In this section, those rules will be discussed and illustrated. Of course, the rules vary according to READ or WRITE operations, but the general problem is this: What happens when the "*w*" part of the A*w* descriptor differs from the number of CHARACTERs declared for the variable? In the examples that follow, we will assume this declaration:

```
CHARACTER*6 HEADER
```

First, let's look at the WRITE cases:

Field	Memory value	Output	Comment
A4	ƀStepƀ	ƀSte	Only the leftmost 4 are output
A6	ƀStepƀ	ƀStepƀ	No problem; same length
A8	ƀStepƀ	ƀƀƀStepƀ	The 6 in memory are right justified in the output

Now, in the case of READ into a CHARACTER∗6 variable:

Field	Input	Memory value	Comments
A4	Amperes	Ampeƀƀ	4 are transferrred, left justified
A6	Amperes	Ampere	6 are transferred
A8	Amperesƀ	peresƀ	Only 6 of the rightmost 8 are transferred

In conclusion, if you omit the "*w*" from the A*w* descriptor, the declared size is used automatically. This is usually just what you want to do, so avoid problems by getting in the habit of omitting the "*w*" part.

7.4 "CHARACTER" "FUNCTIONs"

When FUNCTIONS were discussed in Section 5.1, some details were left out. The same was true for the FORMAT discussion in Chapter 6. This section is intended to fill in the FUNCTION gaps. Let's look at the general case of a CHARACTER FUNCTION declaration:

```
CHARACTER{∗n} FUNCTION function_name ( {arg_list} )
```

where the following definitions apply:

- "*n*" is the length of the CHARACTER string that is returned by the FUNCTION. When the optional "∗*n*" is omitted, only one

CHARACTER is returned by the FUNCTION. There is a special case when "n" is "(*)," that is:

```
CHARACTER*(*) FUNCTION ...
```

This notation means to use the length of the CHARACTER string defined in the calling program—since the FUNCTION *must* be defined in the subprogram. This is illustrated in the example that follows.

- "function_name" is defined in Chapter 5;
- "arg_list" is defined in Chapter 5.

The following example is an outline of a main program, several subprograms, and a FUNCTION that reverses a string. Notice that the FUNCTION is called several times, each time with a different length string. So the FUNCTION must be defined to return strings of differing lengths. Likewise, the length of the string to be reversed is defined by the calling subprogram.

```
        PROGRAM MAINPG
        .
        .
        CHARACTER*10 REVERS
        CHARACTER*10 BACKWD
        .
        .
        BACKWD = REVERS ( 'FORWARD' )
        .
        .
        END
****************
        SUBROUTINE FOLD ...
        .
        .
        CHARACTER*14 REVERS
        CHARACTER*14 OTHWAY
        .
        .
        OTHWAY = REVERS ( OTHWAY )
        .
        .
        END
****************
```

```
      INTEGER FUNCTION UPPER ...
          .
          .
      CHARACTER*5 REVERS
      CHARACTER*5 UPSTRT
      CHARACTER*5 DNSTRT
          .
          .
      DNSTRT = REVERS ( UPSTRT )
          .
          .
      END
*****************
      CHARACTER*(*) FUNCTION REVERS ( STG )
      INTEGER STGLEN
      CHARACTER*(*) STG
          .
          .
      STGLEN = LEN ( STG )
      J = STGLEN
      DO 100 I = 1, STGLEN/2
          REVERS (I:I) = STG (J:J)
          J = J-1
100   CONTINUE
          .
          .
      END
```

Put It Together

Of the two programs that follow, the first deals with the simplified
CHARACTER operations and the second works with CHARACTER
strings and shows how a few of the intrinsic functions work.

The first example contains two "unpacked" CHARACTER arrays,
SCREEN and YLABEL. These represent a graph on a CRT terminal.
YLABEL is an array reserved for Y-axis labels and each element of the
SCREEN array represents a single picture element (pixel). This program
contains examples of several manipulations with these arrays: initializ-

ing, data fill, and printing—so you can use it to get ideas for some of the problems in the exercises. You will probably notice that some pixels are initialized several times—for instance SCREEN(1,1). This is not an oversight; I think this technique makes coding easier and the code more understandable. Also, take time to study the "implied DO" used to display the arrays—it can be fun!

```
      PROGRAM SCREEN
C
C Set up a terminal image, put some junk in it and display it
C
C                        COLS and ROWS are the dimensions of
C                        the graph
      INTEGER COLS, ROWS
      PARAMETER (COLS = 71)
      PARAMETER (ROWS = 21)
C                        SCREEN is the graph data area
      CHARACTER SCREEN (1:COLS, 1:ROWS)
C                        YLABEL is a place to put Y-axis
C                        labels
      CHARACTER YLABEL (1:8, 1:ROWS)

C Initialize the areas
      DO 105 J=1, ROWS
        DO 100 I=1, COLS
          SCREEN (I,J) = ' '
100     CONTINUE
        DO 101 I=1, 8
          YLABEL (I,J) = ' '
101     CONTINUE
105   CONTINUE

C Draw the grids - data, if any, will replace grid marks.
C Intersections are marked with a plus symbol.
      DO 110 I=1, COLS
        DO 110 J=1, ROWS, 5
          SCREEN (I,J) = '_'
110   CONTINUE
      DO 120 I=1, COLS, 10
        DO 120 J=1, ROWS
          SCREEN (I,J) = '¦'
120   CONTINUE
```

```
       DO 130 I=1, COLS, 10
          DO 130 J=1, ROWS, 5
             SCREEN (I,J) = '+'
130     CONTINUE

C Put some data in the graph image
       I = 30
       DO 200 J=1,ROWS
          SCREEN (I,J) = 'X'
          SCREEN (60-I,J) = 'A'
          I=I+1
200     CONTINUE

C Print the graph
       WRITE (*,*) ( (YLABEL (K,J), K=1,8),
     +         (SCREEN (I,J), I=1,COLS), J=1, ROWS )

       END
```

When executed, SCREEN will produce the following display:

```
+----------+----------+--------A+-------- ---------+----------+----------+
¦          ¦          ¦        A X         ¦          ¦          ¦        ¦
¦          ¦          ¦        A ¦X        ¦          ¦          ¦        ¦
¦          ¦          ¦        A ¦ X       ¦          ¦          ¦        ¦
¦          ¦          ¦      A   ¦  X      ¦          ¦          ¦        ¦
+----------+----------+---A-----+---X-----+----------+----------+--------+
¦          ¦          ¦   A      ¦    X     ¦          ¦          ¦        ¦
¦          ¦          ¦  ¦A      ¦    X  ¦          ¦          ¦        ¦
¦          ¦          ¦  ¦A      ¦     X ¦          ¦          ¦        ¦
¦          ¦          ¦   A      ¦      X ¦          ¦          ¦        ¦
+----------+----------+--A------+--------+------X+----------+----------+--------+
¦          ¦          ¦ A ¦      ¦       X          ¦          ¦        ¦
¦          ¦          ¦  A ¦      ¦       ¦X          ¦          ¦        ¦
¦          ¦          ¦  A ¦      ¦       ¦ X         ¦          ¦        ¦
¦          ¦          ¦ A  ¦      ¦       ¦ X         ¦          ¦        ¦
+----------+---A-----+----------+----------+---X-----+----------+--------+
¦          ¦ ¦ A ¦      ¦       ¦      X   ¦          ¦        ¦
¦          ¦ ¦ A ¦      ¦       ¦     X    ¦          ¦        ¦
¦          ¦ ¦A  ¦      ¦       ¦      X ¦          ¦        ¦
¦          ¦ A  ¦      ¦       ¦      X ¦          ¦        ¦
+--------A+----------+----------+----------+--------X+----------+--------+
```

The second program doesn't do much that is interesting. It is intended only to give you a sample of some possibilities. Here are a couple especially worth noting:

a) The use of the PARAMETER MAXCHR in the CHARACTER declarations are in parentheses. This is required.

b) The use of the (∗) in the declaration of variables S, R, and SUBS in the subprograms was discussed in 7.2. It means "use the length defined by the actual variable in the calling program." This notation means that the strings' sizes don't have to be passed as arguments; instead, the programmer can use the "LEN" intrinsic if the subprogram needs to know the defined length.

c) Code in the main program to detect the actual length of the input strings (INPSTG and SCHSTG) could have been identical, but it was done differently to give you some ideas. The best solution would have been to make that code into a subprogram.

Once more, please note: This is one of those programs that just won't run on most micro-based FORTRANs, because substring operations are not included in sub-FORTRAN.

```
      PROGRAM STRING
C
C Prompt for, REVERSe and display a string then
C prompt for a substring, SEARCH for it and display count of hits.
C                         MAXCHR is the maximum number of characters
C                         allowed in the various strings
      INTEGER MAXCHR
      PARAMETER (MAXCHR = 20)

C                         INPSTG is the basic input string
      CHARACTER*(MAXCHR) INPSTG
C                         REVSTG is the reverse of INPSTG
      CHARACTER*(MAXCHR) REVSTG
C                         SCHSTG is the substring to be located
C                         in INPSTG
      CHARACTER*(MAXCHR) SCHSTG
C                         NCHARS is the length of INPSTG
      INTEGER NCHARS
C                         SSLEN is the length of SCHSTG
      INTEGER SSLEN
C                         SEARCH is the FUNCTION name
      INTEGER SEARCH
```

```
C Prompt/read a bunch of characters
100     CONTINUE
        WRITE (*,*) 'Enter a string (use tics)'
        READ (*,*) INPSTG
C
C Compute the actual string length
        NCHARS = MAXCHR
200     CONTINUE
            IF (INPSTG (NCHARS:NCHARS) .EQ. ' ') THEN
                NCHARS = NCHARS - 1
                IF (NCHARS .EQ. 0) THEN
                    WRITE (*,*) 'Really, do something interesting!'
                    GOTO 100
                ENDIF
                GOTO 200
            ENDIF
250     CONTINUE
C
C Reverse and print the result
        CALL REVERS ( INPSTG, NCHARS, REVSTG )
        WRITE (*,*) 'That is', NCHARS,
     +              ' characters long. The reversed string is:'
        WRITE (*,*) REVSTG (1:NCHARS)
C
C Prompt/read a substring used for searching
300     CONTINUE
        WRITE (*,*) 'Enter a substring (use tics)'
        READ (*,*) SCHSTG
C
C Find length of substring
        DO 310 SSLEN = MAXCHR, 1, -1
            IF (SCHSTG (SSLEN:SSLEN) .NE. ' ') THEN
                GOTO 350
            ENDIF
310     CONTINUE
        WRITE (*,*) 'A zero length line is boring'
        GOTO 300
350     CONTINUE
        WRITE (*,*) 'That contains', SSLEN, ' characters.'
C
C Using SEARCH, locate if, and how many times that substring occurs
C in the initial string. This is a FUNCTION, so I can use it below.
```

(continued)

```
      WRITE (*,*) 'Substring occurs',
    +           SEARCH (INPSTG, NCHARS, SCHSTG, SSLEN), ' times.'
C
C Infinite loop
      GOTO 100

      END
C ***********************************************************
C REVERS will reverse the input string (S) and put the
C result in the output string (R)

      SUBROUTINE REVERS ( S, L, R )

C                      S (input) is the input string
      CHARACTER*(*) S
C                      L (input) is the length of S
      INTEGER L
C                      R (output) is the reverse of S
      CHARACTER*(*) R
C                      J points to the right-end of the string
      INTEGER J
C                      I points to the left-end of the string
      INTEGER I

C J must initially be at the right-end
      J = L

C Loop thru half the number of characters.
C I is the left-end pointer.
      DO 100 I = 1, J/2

C Move the left-end of "S" to the right-end of "R" and
      R (J:J) = S (I:I)

C Move the right-end of "S" to the left-end of "R"
      R (I:I) = S (J:J)

C Move the right-end pointer to the left
C The left-end is moved by the DO
      J = J - 1
100   CONTINUE

      END
C ***********************************************************
```

```
C SEARCH counts the number of instances the substring (SUBS) occurs
C in the input string (S)

        INTEGER FUNCTION SEARCH ( S, L, SUBS, SUBL )

C                       S (input) is the string to be searched
        CHARACTER*(*) S
C                       L (input) is the length of S
        INTEGER L
C                       SUBS (input) is the substring
        CHARACTER*(*) SUBS
C                       SUBL (input) is the length of SUBS
        INTEGER SUBL

C                       LEND is the left-end starting point of the
C                       next search on S
        INTEGER LEND
C                       FINDIT is the location of the substring in
C                       the modified version of S
C       INTEGER FINDIT
C
C Start by looking at the entire string
        SEARCH = 0
        LEND = 1
C
C Loop until substring is no longer found
100     CONTINUE
        IF (LEND .LE. L) THEN
C
C Look for substring. INDEX is zero if not found.
            FINDIT = INDEX ( S(LEND:L), SUBS(1:SUBL) )
            IF ( FINDIT .NE. 0 ) THEN
C
C Reset starting point for next look
                LEND = LEND + FINDIT
C
C Count the occurrence
                SEARCH = SEARCH + 1
                GOTO 100
            ELSE
                GOTO 200
            ENDIF
        ENDIF
200     CONTINUE

        END
```

When RUN, STRING will produce the following typical dialog:

```
$ RUN STRINGS
Enter a string (use tics)
"
Really, do something interesting!
Enter a string (use tics)
' '
Really, do something interesting!
Enter a string (use tics)
'do something int int interesting'
That is          20 characters long. The reversed string is:
tni tni gnihtemos od
Enter a substring (use tics)
"
A zero length line is boring
Enter a substring (use tics)
'int'
That contains         3 characters.
Substring occurs          2 times.
Enter a string (use tics)
'(use tics)'
That is          10 characters long. The reversed string is:
)scit esu(
Enter a substring (use tics)
'('
That contains         1 characters.
Substring occurs          1 times.
```

Pitfalls

Because complete CHARACTER variable type operations are available only on full versions of standard FORTRAN 77, you must be certain that the features you desire are actually available on the FORTRAN version on your system. If you don't have a manual handy, just code up the examples to see if they compile without errors. Errors probably indicate that the particular feature doesn't exist on your FORTRAN.

Other general comments that can be made safely are:

■ When doing READ/WRITE operations of CHARACTER variables, use the simple A descriptor rather than the more complicated A*w* form. This eliminates the problems associated with specifying inconsistent length strings.

■ When using a READ (*,*) for CHARACTERS, the user *must* use tic marks (') around the input. If this is too hard to live with, then have the READ reference a FORMAT (A) statement instead.

■ When passing CHARACTER arguments to subprograms, define their length in the subprogram with the "*(*)" form. This eliminates the problems associated with working with strings of differing lengths.

■ When working with the multiargument intrinsics, be sure you are specifying the order of the arguments correctly.

■ Working with arrays of strings is awkward to Pascal programmers because the order of the "subscripts" is backwards from what they know.

■ FORTRAN allows CHARACTER constants to be shorter or longer than the definition; truncation or padding is done automatically. In Pascal you were used to constants that were exactly the same length as the variable they were to be stored in. You will have to get used to the padding and truncation rules in FORTRAN.

Exercises

1. Design/code a point plotter SUBROUTINE for a 22 × 79 grid like the example program. Your program should interface with the user in the following manner:
   a) Have the user enter the number of points and then the actual pairs of points to be plotted.
   b) Scale the data set—that is, figure out what the X- and Y-pixel sizes should be. This is done by finding the maximum and minimum in each direction, using the MAX and MIN intrinsics, and dividing by the number of pixels in that direction.
   c) Internally build the graph and store the data points in it.

d) Display the graph.

e) Loop to (a).

Note that this problem has been simplified as much as possible just to get you into it. For instance, there are no labels on the graph. The following problems will build on this program.

2. Refer to Exercise 1. You probably used an "X" or something similar to plot the data point. And you probably noticed that the Y-axis granularity (only 22 pixels) makes any graph rather "jagged." In this problem, we are offering a suggestion that may smooth out your graph. If you plot data points with the nearly identical characters, minus " − " and the underbar " _ ", you can double the Y-pixels. Since the minus is in the middle of the line and the underbar is at the bottom, each row of the display can be divided in half, so you really have 44 pixels in the Y-direction.

   Modify Exercise 1 to scale for a 44 × 79 graph, even though it still has 22 rows in which to be displayed, and then to plot with either a minus or an underbar in the appropriate row. One problem you will run into is that the graph borders and grid lines are made with minuses. You might want to use something else so that the data stands out from the lines.

3. Redesign either Exercise 1 or 2 above to include labeling and numbering of the axes. Labeling takes room on the screen, so pixels must be sacrificed. Good engineering practice presents only numbers which are multiples of 1, 2, or 5 on the axes.

4. Redesign Exercise 1, 2, or 3 to include connecting the data points with "straight" lines. This is tricky, because the data must be sorted before they can be plotted, and extra data points must be computed. Try this algorithm:

   a) Sort the X-data points, moving the corresponding Y-data points as the X-data move.

   b) Find the Y minimum and maximum. The X values are known through the sort operation.

   c) Scale the X and Y axes.

   d) Enter the leftmost data point into the graph data area.

   e) Enter the next data point into the graph data area.

   f) Compute the coefficients of the straight line connecting the last two points plotted.

g) For each empty X-pixel between the last two points, use the straight line equation to find the corresponding Y-pixel to be filled. Enter a data point in the graph area for the line too.

h) Repeat steps (e) through (g) for the remaining data points.

i) Plot the resulting graph.

5. So far we've looked only at two-dimensional graphing. Three-dimensional graphing of solids (not lines) is something that you can do too. These are called "contour plots." The Z-axis is represented by plotting with a digit, not an X. This means that the Z-axis has only 11 pixels to work with: blank, 0, 1, 2, . . ., 9. You will see that this is only a coarse representation of a solid, but it is still a worthwhile tool. It is possible to input data points the same way the previous problems were structured, but that is a bit awkward. Instead, you will plot a "surface" represented by an equation which is built into your program. Defining the limits of the graph is also a little awkward. We are not going to do numerical differentiation, but rather let the user set the limits. This allows the user to get a quick and easy overview of the function. Here is the algorithm:

a) Prompt/read the X-, Y-, and Z-axis lower and upper limits.

b) Using the input limits, scale the X-, Y-, and Z- pixels.

c) For *every* pixel in the X-Y plane, evaluate the equation below to find Z. If that Z is within the specified limits, scale it and enter the appropriate digit into the graph data area. Otherwise that pixel is left blank.

d) Print the graph data area.

e) Loop to step (a).

Here are some equations you may want to try out:

a) $Z = X^2 + Y^2 + 4X - 6Y$

b) $Z = \sqrt{X^2 + 4Y^2 - 8Y}$

c) $Z = 4XY$

d) $Z = e^{x-y} \, \text{Sin}(4x) \, \text{Cos}(3y)$

6. If you liked Exercise 5, you will probably be intrigued by this one. Although mathematically correct, plotting a single digit isn't very user friendly—that is, the display still requires a lot of interpretation. In this problem, you are to use overprinting to represent the Z-axis. You can use a dark blob to represent either end of the scale, and you will have to develop some "gray"

intermediate representations too. Here are some possible overprint suggestions; see if you like them:

Printed symbol	Overprinted characters
▮	H S W M I
И	W M
B	H S
ä	@ #
∦	+ #
∗	∗ +
▬	= –

This is probably not a complete "gray scale," but it does give you a jumping-off point; you will need to experiment with this. Your printer will, no doubt, do some things that mine can't, so you have to adapt your output to your printer.

You will also need to give some thought to a data representation. A three-dimensional data area may be a solution; think of the third dimension as the overprint characters. You will need one layer for each character in the overprint—five layers if you use the above suggestion—and fill layers of the data area instead of entering a digit. When you print, one line of the graph will be made by overprinting the layers. The algorithm in Exercise 5 is modified as follows:

a) For *every* pixel in the X-Y plane, evaluate the equation to find Z. If that Z is within the specified limits, scale it and enter the necessary overprint characters into the graph data area. Otherwise that pixel is left blank.

b) Print one line of the first layer of the data area with a normal carriage control character (a space) and then print the corresponding line of each of the remaining layers with the overprint carriage control character (+). Repeat for all lines in the graph.

Once you get this program running, it will produce a much nicer plot than Exercise 5 does. By the way, you can use the equations from that problem again if you like.

7. Another form of plotting is histograms, or bar charts. These plots represent discrete rather than continuous functions. Normally, this means that scaling is much easier. A histogram can be plotted either horizontally or vertically. We will leave the horizontal discussion until the next exercise. Here is a vertical example; it is the distribution of grades in one of my classes:

```
                         X
              X          XX
              X     X  X XX              X
 X            X   X XX X XX XXX     XX XX X    XX          X   X
+---------+---------+---------+---------+---------+---------+---------+
     90        100       110       120       130       140       150
```

SCORE

You'll note that four students have a score of 110, three have a score of 100 and one has a score of 138.

This program is much like Exercise 1. Develop a program to display a vertical histogram on the screen—a 22 × 79 area. Here is the algorithm you should develop:

a) Have the user enter the number of points, then the actual values to be plotted. In this case, a "point" is represented only by a single integer.

b) Count the data and scale it, if necessary, so that the histogram isn't too high. Since you are now working with counts, an X may represent one or more data values. The horizontal axis should likewise be integer scaled to fit on a single line.

c) Internally build the histogram and store the data points in it.

d) Display the histogram.

e) Loop to (a).

8. Create a horizontal histogram by reworking Exercise 7.

9. Look at the example program in "Put It Together." Why did I pick those particular screen dimensions? Copy that code, and manipulate the dimensions to see what happens. Write a brief paper to justify either:

a) the dimensions I selected, or

b) a better set of dimensions.

10. Redesign any of Exercises 1 through 6 to read data from a file instead of the terminal. The user should be allowed to enter the file's name from the terminal. You have all the basic knowledge to do this problem. However, should you get stuck on the mechanics of READing from a file, look at Chapter 10.

11. Redesign any of Exercises 1 through 5 for a printer instead of the terminal. Printers typically have 132 or 136 columns and 66 or 88 lines, so the granularity of your plot becomes finer. You will have to research your printer to find out its specifications.

12. Combine all the capabilities of Exercises 10 and 11 into one gigantic program. Have your program dialog with the user to find out which options are desired: terminal or file input, terminal or printer output. You should offer the user the option of reviewing the graph on the terminal and then, if it is ok, optionally printing it.

"COMPLEX" OPERATIONS

Here is a real special case, even more so than CHARACTER operations. The topic of COMPLEX variables could have been included with the other variable types, but there are many differences that you should be aware of; much of the time they don't behave like regular variables. In short, this topic deserves a chapter of its own so that all of the strangeness of dealing with COMPLEX variables can be put in one place for you to see and experiment with, and so that you have it for future reference purposes.

Since COMPLEX number applications turn up most often in the study of electronic and electrical applications, many of the exercises deal with this subject. The student who has mastered AC circuit analysis will be

impressed, because COMPLEX arithmetic is easily programmed in FOR-
TRAN and I think he or she will be much happier working with FOR-
TRAN than with paper and pencil. Those unfamiliar with electrical engineering topics shouldn't be discouraged because all necessary
introductory material has been included in the problems. Besides, some
of the problems are more general in nature.

8.1 "COMPLEX" VARIABLES

COMPLEX variables are declared like other variables. Some FORTRANs
even allow DOUBLE PRECISION COMPLEX variables. Here are some
examples:

```
COMPLEX TLINE
COMPLEX LVOLTS(100)
COMPLEX*16 LAMPS(45)
```

The first declaration is for a simple COMPLEX variable, TLINE. The second declaration creates a COMPLEX array, 100 elements long, called
LVOLTS. The third example makes a DOUBLE PRECISION COMPLEX
array, 45 elements long. This third form is not standard but is acceptable
to both VAX® and MS-FORTRAN®. This is all I want to say about DOU-
BLE PRECISION COMPLEX variables, because the details are very specific depending on the implementation and, for many compilers, difficult
to use.

COMPLEX data are actually stored internally as two REAL numbers,
so when you have an expresison like:

```
LVOLTS(I) = TLINE
```

two REAL numbers are moved automatically from TLINE into LVOLTS.
This is more easily seen when you initialize a variable to a constant:

```
TLINE = ( 1.3, 0.0045 )
```

The real part is 1.3 and the imaginary part is .0045. Notice that an
"i"/"j" is not appended to the imaginary part. Also notice that the parentheses and the comma are a necessary part of the constant syntax. You
are allowed to use "E" notation for either or both parts of the COMPLEX
constant:

```
PROPA = ( 0.67, 0.93E-3 )
```

This form of the constant is expected when you READ (∗,∗) a COMPLEX variable from the terminal.

The arithmetic operations, "+", "−", "∗", "/" and "∗∗" are all defined for COMPLEX variables, so expressions like this one are quite valid:

```
TLINE = ( TLINE**2.0 + (1.0, -1.0) ) / LVOLTS ( J )
```

The above equation squares TLINE, adds a COMPLEX constant $(1, -1)$ to that, then divides that sum by an element from the LVOLTS array.

In addition, several of the intrinsics you already know and have used are also defined for COMPLEX variables:

complex	= COS(complex)	Compute the COMPLEX cosine of a COMPLEX number
complex	= EXP(complex)	Raise "e" to a COMPLEX power
complex	= LOG(complex)	Compute the COMPLEX natural logarithm of a COMPLEX number
complex	= SIN(complex)	Compute the COMPLEX sine of a COMPLEX number
real	= SQRT (complex)	Compute the REAL square root of a COMPLEX number; notice that this function doesn't yield a COMPLEX result like the others do

Besides these familiar operations and functions, there are several other built-in functions that are designed especially to deal with COMPLEX numbers. Here is a table of these functions:

real	= ABS(complex)	Computes the magnitude of a COMPLEX number and returns it as a REAL number
real	= AIMAG (complex)	Returns the imaginary part of a COMPLEX number as a REAL number
complex	= CMPLX (real,real)	Creates a COMPLEX number from two REAL numbers
complex	= CONJG (complex)	Returns the COMPLEX conjugate of a COMPLEX number
real	= REAL (complex)	Returns the real part of a COMPLEX number as a REAL number

Notice especially that ABS and REAL have a very special purpose when used with COMPLEX variables.

Parenthetically, here is a common application of the intrinsics. It is often convenient to express imaginary numbers in polar form, that is, magnitude and angle. And so the following equations are for switching between the two systems:

Cartesian (z) to polar (r, alpha)

$$r = ABS (z)$$

$$alpha = ATAN2 (AIMAG (z), REAL (z))$$

Polar (r, alpha) to Cartesian (z)

$$z = CMPLX (r*COS (alpha), r*SIN (alpha))$$

But there are some exceptions—intrinsics that are *not* supplied in the standard FORTRAN 77 library. Certain functions should be underlined as missing, because you may expect to see them without bothering to check first. So, when you code your special program and find it either won't compile or gives unexpected results, look at the list that follows; it contains the built-in functions you could reasonably expect to be defined for COMPLEX variables, but aren't:

ACOS	Arccosine
ASIN	Arcsine
ATAN	Arctangent
COSH	Hyperbolic cosine
SINH	Hyperbolic sine
TAN	Tangent
TANH	Hyperbolic tangent

There is really no big loss here; you can easily "manufacture" these functions, and you will have an opportunity to create the hyperbolic functions in the Exercises.

One final topic—comparing COMPLEX numbers. Is (5., 5.) larger than (−10., −10.)? Certainly not in magnitude, but what is the relational rule? Your best bet is to create your own functions to compare COMPLEX numbers, according to the definition that suits your purposes. FORTRAN 77 will only support .EQ. and .NE. of COMPLEX numbers; both real parts and both imaginary parts must be equal for a COMPLEX number to be equal. But we have discussed the problems you could run into when you compare REAL numbers for equality or inequality, so it is best to stay clear of .EQ. and .NE.. You will find an exercise dealing further with this subject.

8.2 "COMPLEX" I/O

Doing I/O with COMPLEX variables can be confusing, and this section is intended to clarify your thoughts. First of all, when not using a FORMAT, the WRITE (*,*) statement will generate two real numbers, complete with parentheses and comma, just like the COMPLEX constant. The READ (*,*) statement will demand that you enter a COMPLEX constant, complete with parentheses and comma.

When READing and WRITEing COMPLEX numbers using a FORMAT statement, the use of the parentheses and comma is entirely your option. But you must keep in mind that you are working with *two* REAL numbers for each COMPLEX variable. That means you must include two REAL number descriptors for each COMPLEX number you expect to input or output, because there is no special descriptor for COMPLEX variables. Here is an example:

```
      INTEGER LOOP
      COMPLEX ZZERO
          .
          .

      WRITE (*, 10100) LOOP, ZZERO
10100 FORMAT (I5, 2F10.3)
          .
```

The point of this example is that there are only *two* variables in the I/O list but there are *three* descriptors in the FORMAT statement; two are needed for ZZERO because it is COMPLEX. Also notice that the values associated with ZZERO are to be displayed in the standard F form and will not include parentheses or the comma. You may also use the E descriptor for COMPLEX variables. In fact, you can use the F for one part and E for the other, if you like.

Put It Together

This problem will be unfamiliar to you if you're not an Electrical Engineering student. If that is the case, think of it as merely evaluating a couple of complex algebraic equations. On the other hand, even if you have never thought about transmission lines, this is an interesting problem because it applies many COMPLEX number concepts.

Transmission line applications are found in power lines, in telephone lines, in some radar applications, and even in computer hardware interconnections. Whenever alternating current (AC) is applied to a pair of wires, transmission line theory can explain how voltages and currents behave in the wires. This problem will look at two aspects of the theory. You can skip the explanations if you like, and move directly to the equations, but we encourage you to expand your mind.

The first thing to recognize is that a pair of wires has a certain electrical characteristic—called the "characteristic impedance." The wires are really very small inductors (L) and there is a certain resistance (R) in the wire. Additionally, there is a capacitance (C) between the wires, and a certain amount of leakage resistance (G) between them. The resistances are constants, but inductance and capacitance depend on frequency. It shouldn't be surprising that the characteristic impedance (Z_0) can be expressed in complex arithmetic:

$$Z_0 = \sqrt{\frac{R + j\omega L}{G + j\omega C}}$$

where

$$\omega = 2\pi f$$

Nothing is simple, and this is no exception. Another convenient concept in transmission line theory looks very similar. This is called the propagation constant (γ). It too is a complex expression.

$$\gamma = \sqrt{(R + j\omega L)(G + j\omega C)}$$

Note that electrical engineers use "j" instead of "i" to denote the imaginary part of a complex number.

In general terms, this function determines how the current and voltage "die out" over the length of the line. Have you ever noticed that, when you jerk a rope up and down quickly, the "jerk" you inserted travels toward the other end at a certain rate of speed, and the height of the "jerk" gets smaller as it travels? The analogous thing takes place in a transmission line; the characteristic impedance and propagation constant are what govern what comes out the other end of the pair of wires.

While you're not an expert in transmission line theory, you know a little more now. It is time to look at the program. The two equations above have been coded into it, the equations for different values of frequency have been evaluated, and the answers have been displayed. It is relatively easy, even if you didn't understand the theory. This is the algorithm:

1. Initialize a loop to be executed 11 times
   a) Compute "frequency" and "omega."
   b) Compute "characteristic impedance" and express it in terms of magnitude and angle.
   c) Compute "propagation constant" and express it in terms of magnitude and angle.
   d) Display the results.
2. Done.

```
      PROGRAM ZZERO
C
C Investigate how the Characteristic Impedance and Propagation
C Constant of a transmission line vary with frequency.
C
C                    R, L C and G are line constants
      REAL R
      REAL L
      REAL C
      REAL G
C                    IMPED is the characteristic impedance
      COMPLEX IMPED
C                    PROPA is the propagation constant
      COMPLEX PROPA
C                    LOOPER is the loop variable
      INTEGER LOOPER
                     OMEGA, FREQ are frequency variables
      REAL OMEGA
      REAL FREQ
C                    PEDMAG and PROMAG are IMPED and PROPA magnitude
      REAL PEDMAG
      REAL PROMAG
                     PEDANG and PROANG are IMPED and PROPA angles
      REAL PEDANG
      REAL PROANG
C                    TWOPI is the constant 2 * pi
      REAL TWOPI
C                    RAD2DG is radians to degrees constant
      REAL RAD2DG
```

(continued)

```
C
C                              INITIALIZED CONSTANTS
       DATA TWOPI /6.28318531/
       DATA R /10.2/
       DATA L /.00367/
       DATA C /.00821E-6/
       DATA G /.3E-6/
       DATA RAD2DG /57.2957795/
C
C Print out the table header
       WRITE (*,10000) 'Freq', 'Char Imped', 'Magnit', 'Phase',
      +                         'Prop Const', 'Magnit', 'Phase'
10000  FORMAT ( 3X, A, 3(4X, A), ' ¦ ', 3(3X, A) )
       WRITE (*,10000) ' Hz ', 'Real Imag', ' Ohms ', 'Deg. '
      +                         'Real Imag', ' Ohms ', 'Deg. '
       WRITE (*,10001)
10001  FORMAT ( 1X, 78('-') )
C
C Loop through the decades - looking for general shape of curve
       DO 200 LOOPER = 1, 11
C
C          Compute frequency - 2 data points per decade
           FREQ = 10.0**( (LOOPER-1.0)/2.0 )
           OMEGA = FREQ * TWOPI
C
C          Compute Characteristic Impedance stuff
           IMPED = SQRT ( CMPLX (R, OMEGA*L) / CMPLX (G, OMEGA*C) )
           PEDMAG = SQRT ( REAL(IMPED)**2 + AIMAG(IMPED)**2 )
           PEDANG = ATAN2 ( AIMAG(IMPED), REAL(IMPED) ) * RAD2DG
C
C          Compute Propagation Constant stuff
           PROPA = SQRT ( CMPLX (R, OMEGA*L) * CMPLX (G, OMEGA*C) )
           PROMAG = ABS (PROPA)
           PROANG = ATAN2 ( AIMAG(PROPA), REAL(PROPA) ) * RAD2DG
C
C          Display the results
           WRITE (*,20000) FREQ, IMPED, PEDMAG, PEDANG,
      +                         PROPA, PROMAG, PROANG
20000      FORMAT (F8.0, 2F7.0, F8.0, F9.2, '  ¦',
      +                         2F7.4, F8.4, F8.2 )
200    CONTINUE

       END
```

When this program is run, the following is produced:

Freq Hz	Char Real	Imped Imag	Magnit Ohms	Phase Deg.		Prop Real	Const Imag	Magnit Ohms	Phase Deg.
1.	5768.	−486.	5789.	−4.81		.0018	.0002	.0018	4.94
3.	5302.	−1328.	5465.	−14.06		.0018	.0005	.0019	14.47
10.	3607.	−2021.	4135.	−29.26		.0021	.0013	.0025	30.56
32.	1963.	−1520.	2483.	−37.75		.0031	.0027	.0041	41.83
100.	1137.	−855.	1423.	−36.97		.0048	.0056	.0074	49.71
316.	783.	−394.	877.	−26.69		.0067	.0127	.0143	62.25
1000.	684.	−143.	699.	−11.76		.0076	.0353	.0361	77.90
3162.	670.	−46.	672.	−3.93		.0077	.1093	.1096	85.97
10000.	669.	−15.	669.	−1.25		.0077	.3450	.3451	88.72
31623.	669.	−5.	669.	−.40		.0077	1.0907	1.0907	89.59
100000.	669.	−1.	669.	−.13		.0077	3.4489	3.4489	89.87

Pitfalls

Probably the major mistakes you make will be using the FORMAT with COMPLEX variables. Remember, you must have two specifications for each COMPLEX variable in the I/O list. There is also a tendency to forget to include the parentheses and comma when working with the READ (*,*) or with constants. Since there are two descriptors, two numbers will be displayed in an output listing, so be sure to allow enough room for them.

Another item not specifically mentioned in the text, but which may cause you to pause is that when you have a REAL and a COMPLEX in an expression, the result is COMPLEX. The same is true with INTEGER and COMPLEX. FORTRAN will take care of the conversions for you properly, but you must remember what is happening.

A final word of caution in using intrinsics: Not all intrinsics have been defined to work properly with COMPLEX variables, even though there doesn't seem to be a good reason for the restriction. Be very suspicious of errors you get when working with built-in functions.

Exercises

1. Since COMPLEX arithmetic aids (like calculators and tables) are generally unavailable, it would be convenient to have such a tool for the other problems in this chapter. In this problem you will develop a sort of on-line COMPLEX number cruncher. This is the algorithm your program should perform:
a) Prompt/read two COMPLEX numbers.
b) Prompt/read an operation: "+", "−", "*", or "/".
c) Perform the indicated operation on the two COMPLEX numbers.
d) Display the result in COMPLEX form.
e) Display the result in magnitude and angle form.
f) Loop to step (a)

2. This is not even a COMPLEX problem; this is simply a good place to ask the question. Why are there two arctangent functions? Compare the results of these two functions (ATAN and ATAN2) with arguments in all four quadrants—that is (x,y), $(-x,y)$, $(-x,-y)$, and $(x,-y)$. Write a program that displays a table like the following:

```
  X  ¦  Y  ¦   ATAN (X/Y)   ¦   ATAN2 (X, Y)
-----+-----+---------------+----------------
```

Based on the results presented in your table, write a short paper describing your observations. Obviously there is a problem here, or we wouldn't have to bother writing a program. Why does the problem occur with the arctangent function and not with the arcsine or arccosine functions? Hint: Compare the graphs of these three functions. What is peculiar about the arctangent function?

3. The COMPLEX arithmetic supported by FORTRAN isn't limited to complex numbers. This feature can be applied to two-dimensional vector arithmetic also. Like Exercise 1, arithmetic aids are generally unavailable to do vector manipulations, so this problem will create such a tool. The algorithm you are to implement is:
a) Prompt/read a vector in polar form: magnitude and angle.
b) Prompt/read a second vector.
c) Convert the two vectors into COMPLEX numbers.

d) Prompt/read an operation: "+" or "−".
e) Display the result in polar form.
f) Loop to step (a).

4. This is a problem in COMPLEX algebra. The functions in this problem are not defined in FORTRAN 77 for COMPLEX variables. So, create FUNCTIONs for TAN, SINH, COSH, and TANH for the COMPLEX argument, "z." The equations for these functions are:

TAN(z)   = SIN(z)/COS(z)
SINH(z) = $(e^z - e^{-z})/2$
COSH(z) = $(e^z + e^{-z})/2$
TANH(z) = SINH(z)/COSH(z)

Of course the code is trivial, but you have to check your answers anyway. I know of no calculator that does COMPLEX arithmetic, so the only definitive source available is in your library. The National Bureau of Standards publishes such a table.

5. Refer to your contour plotter developed in the last chapter. From previous class work, or perhaps even high school, you have a good feeling for the behavior of SIN and COS for REAL arguments. But what do these functions look like in the COMPLEX plane? There are many contour plots that could be made; the possibilities are listed below. Design a modified contour plotter to present one or more of these graphs. Make the X-axis the real input argument and the Y-axis the imaginary input argument. The Z-axis is one out of the following list:
a) Z = ¦ SIN ¦
b) Z = real part (SIN)
c) Z = imaginary part (SIN)
d) Z = Phase angle (SIN)
e) Look at (a) through (d) for COS instead of SIN. Is it simply a shifted version of SIN?

6. You have probably never done anything with the hyperbolic functions, much less the hyperbolic functions with COMPLEX arguments, so you don't have any feeling at all of how these functions behave. Exercise 4 gives you the hyperbolic expressions you will have to define for the computational portion of this exercise, and Exercise 5 gives you some thoughts on the graphic presentation of these functions. In fact, you have all the same options presented in Exercise 5: Plot the input COMPLEX argument in the X-Y plane, and the Z-axis can be any of the following:

a) Z = ¦ SINH ¦
b) Z = real part (SINH)
c) Z = imaginary part (SINH)
d) Z = phase angle (SINH)
e) (a) through (d) using the COSH instead
f) (a) through (d) using the TANH instead

7. You are to investigate the performance of the two circuits below. If you have had an AC circuit analysis course, you will immediately recognize them as "series and parallel resonant" circuits, using an inductor (or coil, L) and a capacitor (C). The presence of the resistor (R) in the circuit is the subject of the next problem. You will design/code a model of these circuits and investigate how they behave when the frequency (f) is varied and (in Exercise 8) when the resistance is varied.

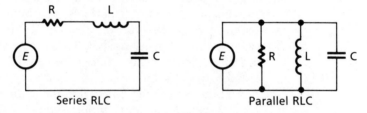

Series RLC Parallel RLC

Let's get some preliminaries out of the way. The resonant frequency (f_o) of a series LC circuit is given by:

$$f_o = \frac{1}{2\pi\sqrt{LC}}$$

The impedance of an inductor (Z_1) in terms of complex numbers is:

$$Z_1 = (0.0, 2\pi f L)$$

And the impedance of a capacitor (Z_c) in terms of complex numbers is:

$$Z_c = \left(0.0, \frac{-1}{2\pi f C}\right)$$

Perhaps you still remember what the current (I) flowing in a circuit having an impedance (Z) supplied with a voltage (E) is.

And, it shouldn't be surprising to find out that the current is a complex number if the impedance is; it doesn't matter if the voltage is complex or not.

$$I = E / Z$$

Note also that impedances in series simply add together. And impedances in parallel add according to their inverses. This means that the following equation gives the equivalent impedance (Z_{eq}) of two impedances in parallel:

$$Z_{eq} = \frac{1}{\dfrac{1}{Z_1} + \dfrac{1}{Z_2}} = \frac{Z_1 Z_2}{Z_1 + Z_2}$$

Now, we are ready to state the problem. Design/code a program to fill in the following table. Use these values when evaluating the equations: $E = 10$ volts, $R = 25$ ohms, $L = .005$ henries, $C = 2 \times 10^{-6}$ farads.

f	Parallel RLC		Series RLC	
	REAL (I)	Phase (I)	REAL (I)	Phase (I)
1000				
1100				
1200				
.				
.				
1900				
2000				

8. Referring to the previous problem, examine the effect of the value of the resistance (R) on the circuit. Rework your program to create several tables like the one for Exercise 7: one for each of these values of R: 1, 10, 100, 500. If you have a plotter package working, figure out a creative way to present all these data. You will probably want to separate the Parallel and the Series information. If you don't plot the data, write a short paper describing the results of varying R. Generally, R represents a "goodness" measure: The smaller R is, the better.

9. If you get this one running, you can probably sell it to your Electrical Engineering Department. Examine sending impedance (Z_s), voltage (E), current (I) and power (P) along a length of a lossy transmission line. A picture which shows the notation might be a help—if you can read schematics:

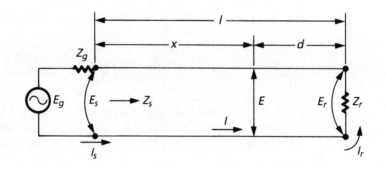

The purpose behind all of this is to stuff voltage in at the left and hope to see something come out on the right. E_g and Z_g are the measurable voltage and impedance of the generator—thus they are constants. E_g and Z_g are real. Z_0 is the characteristic impedance of the transmission line and it is a measured complex constant.

Additionally, this line has a propagation constant, "γ," associated with it, and it is complex too. The transmission line we are dealing with is "l" miles long. The final constant in this system is the complex impedance at the receiving end of the transmission line, Z_r, and it is a constant too: It represents the load. The remaining quantities are computed in terms of these constants.

The computational algorithm follows. The first of the calculated variables in this system is Z_s. It is the impedance that the generator "sees" at the sending end of the transmission line and it varies with the length of the transmission line. Once Z_s is computed, I_s and E_s can be found too. Of course, since Z_s varies with length, so will I_s and E_s.

$$Z_s = Z_0 \frac{Z_r + Z_0 \tanh (\gamma l)}{Z_0 + Z_r \tanh (\gamma l)}$$

$$I_s = E_g / (Z_g + Z_s) \quad \text{and} \quad E_s = I_s Z_s$$

The guts of this problem is to examine the variation of voltage and current along the length of the line. E is the complex voltage and I is the complex current at some point "x" which is measured from the sending end—the left end. When "x" becomes "l"—that is, moving to the right on the line—in the equations below, then E becomes E_r and I becomes I_r—the receiving end voltage and current. So, general terms, Z, E, and I at some point "x" are:

$$Z = Z_0 \frac{Z_s + Z_0 \tanh (\gamma x)}{Z_0 + Z_s \tanh (\gamma x)}$$

$$E = E_s \left(\cosh (\gamma x) - \frac{Z_0}{Z_s} \sinh (\gamma x)\right)$$

$$I = I_s \left(\cosh (\gamma x) - \frac{Z_s}{Z_0} \sinh (\gamma x)\right)$$

P, the power in the line at point "x," can be computed several different ways once the above are known. The easiest may look familiar from high school physics:

$$P = |E| \, |I|$$

Now, using the expressions from the "Put It Together" example, a frequency (f) of 1000 hertz and a line length of 250 miles, look to see what happens to the voltage, current, and power along the line. Prepare a table in 5-mile increments (51 entries) vertically, and E, I, and P horizontally.

10. Plot results of Exercise 9. The usual method is to assign wavelength to the X-axis—this "normalizes" the frequency—and one of the following to the Y-axis:
a) $|Z_s|$
b) phase angle of Z_s
c) $|E|$
d) $|I|$
e) $|P|$
To convert from miles (l) to wavelength (λ), use the following equation:

$$\lambda = lf \, / \, 186282.397$$

which is, in our case, with $f = 1000$:

$$\lambda = l \, / \, 186.282397$$

11. This problem involves trivial logic for COMPLEX numbers. Design/code three FUNCTIONs that compare COMPLEX numbers:

INTEGER FUNCTION COMCOM (complex_arg_1, complex_arg_2)

a) Convert the points to a magnitude, then do a comparison; return an integer to signal the result:

-1 for .LT.

0 for .EQ.

1 for .GT.

b) Compare by quadrant—where a point in quadrant IV > III > II > I. If both points lie in the same quadrant or are between the same two quadrants—that is, on the same axis—they are equal. Return the same integer codes as part (a).

c) Compare imaginary parts only—that is, a point above the X-axis is larger than a point below the X-axis. If both points are above, on, or below the X-axis, they are equal. Use the same codes as part (a).

9

CREATING A SYSTEM OF FORTRAN PROGRAMS

Up to this point, we have only investigated passing data between subprograms through the argument lists; we have not introduced any method for creating "global" variables in a FORTRAN program. Yet, as mentioned earlier, there is another method for sharing variables. As the name implies, that construct is COMMON. It is needed only for larger systems of programs, while arguments work nicely for the smaller problems we've examined up to this point. This chapter is devoted to a detailed examination of COMMON.

9.1 NAMED "COMMON"

COMMON is a mechanism devised to group variables as a "global" unit so they can be referenced by more than one subprogram. Sharing data is

181

what we need to be able to accomplish, and we want to do it without using argument lists. The general form of the COMMON construct is as follows:

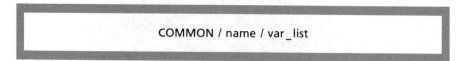

COMMON / name / var_list

where the following definitions apply:

- "name" is a unique label, following the variable name rules, by which this COMMON block is identified. It has no meaning to the programmer other than to help him or her logically organize a program, but it is required by the system to load the program into memory.
- "var_list" is a list of previously declared variable names to be located within the logical block of COMMON.

In order to give you a sense for an organization that can be produced by the use of COMMON, I have created the example system of subprograms that follows. On the left, a FORTRAN program has been symbolized with three subprograms. On the right, two COMMON blocks contain variables used by the program. Variables in the top COMMON block, called "X," are used by the Main Program, by Subprogram A, and by Subprogram C. The variables in the lower COMMON block, "Y," are used only by Subprogram A and Subprogram B.

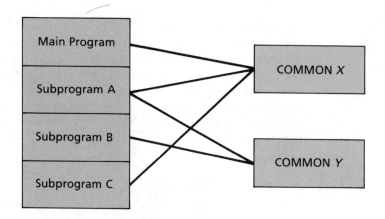

In order to create such a structure, the following code—in skeleton form—shows the essential ingredients. The following guidelines apply:

1. Every subprogram that references a variable in a COMMON block must *redeclare* all the variables in that block—even if all the variables are not used in that subprogram.
2. The COMMON declaration "var_list" must be stated in the *same order* every time it is used.
3. A good programming practice is to group all the variables for the COMMON declaration together in the subprogram. This makes it easier to duplicate the declaration when moving it to another subprogram; it also makes it easier to find the declarations themselves.
4. A COMMON block needs to be documented only once. Duplicate comments are difficult to maintain and make the listing unnecessarily long.

Here is the skeleton code for the diagram:

```
      PROGRAM MAIN
      .
      .
C <> <> <> <> <> <> <> <> <> <> <> <> <> <> <> <>
C COMMON DATA REGION  'X'
      .
      . declarations and documentation for variables in 'X'
      .

      COMMON / X /  list of variables in X
C
      .
      . remainder of 'MAIN'
      .

      END


*****************************************************
C
C Subprogram A description
C
      SUBROUTINE A ( arguments )
```

```
C <> <> <> <> <> <> <> <> <> <> <> <> <> <> <> <>
C COMMON DATA REGION  'X'
      .
      . declarations for variables in 'X'
      .

      COMMON / X /  list of variables in X
C <> <> <> <> <> <> <> <> <> <> <> <> <> <> <> <>
C COMMON DATA REGION  'Y'
      .
      . declarations and documentation for variables in 'Y'
      .

      COMMON / Y /  list of variables in Y
C

      .
      . remainder of 'A'
      .

      END
****************************************************
C
C Subprogram B description
C
      FUNCTION B ( arguments )

C <> <> <> <> <> <> <> <> <> <> <> <> <> <> <> <>
C COMMON DATA REGION  'Y'
      .
      . declarations for variables in 'Y'
      .

      COMMON / Y /  list of variables in Y
C
      .
      . remainder of 'B'
      .

      END
****************************************************
C
C Subprogram C description
```

```
C
      SUBROUTINE C ( arguments )

C <> <> <> <> <> <> <> <> <> <> <> <> <> <> <> <>
C COMMON DATA REGION   'X'
      .
      . declarations for variables in 'X'
      .
      COMMON / X /   list of variables in X
C
      .
      . remainder of 'C'
      .
      END
```

9.2 INITIALIZING "COMMON" VARIABLES

One particular construct has been created just so you can initialize COMMON variables. It is called BLOCK DATA. Actually, you must create something like a subprogram containing only declarations and COMMON statements, with the addition of the familiar DATA statement. The general form of a BLOCK DATA "subprogram" is:

```
BLOCK DATA name

. declarations
.
COMMON / . . .
.
.
DATA . . .
.
.
END
```

The BLOCK DATA subprogram contains the PARAMETER statements and type statements along with the COMMON statement, just as one would in a standard subprogram. In addition, the DATA statements for the variables are included. The BLOCK DATA subprogram may contain multiple COMMON block declarations and initializations. A good programming practice is to limit your system to only one BLOCK DATA subprogram, so that you know where to find *all* COMMON initializations.

Although there are no restrictions in the FORTRAN 77 standard, the actual location of the BLOCK DATA subprogram in the source file may be crucial. For instance, MS-FORTRAN® BLOCK DATA must be *first* in the file; if not, the initializations are simply not picked up—there are no error messages.

Another skeleton program follows. Like the previous one, it has two COMMON blocks ("FIRST" and "SECOND"), two subprograms ("ONE" and "TWO"), and a Main program. The new feature is the BLOCK DATA program, named "COMDAT." In this case, I excused myself from the documentation rules I have been using and put all the comments in the BLOCK DATA program. It is more "natural" to expect initialization and comments together. Also notice that only some of the COMMON variables were initialized. It is not necessary to initialize them all, but you may find it to be a good habit.

```
          BLOCK DATA COMDAT
C
C Initialization of COMMON
C
C <> <> <> <> <> <> <> <> <> <> <> <> <> <> <> <>
C Comments for FIRST
          REAL X, Y
          INTEGER A
          COMMON /FIRST/ A, X, Y
C <> <> <> <> <> <> <> <> <> <> <> <> <> <> <> <>
C Comments for SECOND
          REAL Z
          CHARACTER*2 B
          COMMON /SECOND/ B, Z
C <> <> <> <> <> <> <> <> <> <> <> <> <> <> <> <>
C DATA for FIRST
          DATA A /100/
          DATA Y /-100.0/
C <> <> <> <> <> <> <> <> <> <> <> <> <> <> <> <>
```

```
C DATA for SECOND
        DATA B / 'mw' /

        END


****************************************************
C***************************************************
        PROGRAM COMMON
C
C COMMON examples with BLOCK DATA
C
        REAL X, Y
        INTEGER A
        COMMON /FIRST/ A, X, Y

        REAL Z
        CHARACTER*2 B
        COMMON /SECOND/ B, Z

        CALL ONE
        CALL TWO
C       .
C       . rest of Main Program
C       .

        END

****************************************************
        SUBROUTINE ONE

        REAL X, Y
        INTEGER A
        COMMON /FIRST/ A, X, Y

        WRITE (*,*) 'From ONE:', A, X, Y
C       .
C       . rest of ONE
C       .

        END

****************************************************
```

(continued)

```
      SUBROUTINE TWO

      REAL Z
      CHARACTER*2 B
      COMMON /SECOND/ B, Z

      WRITE (*,*) 'From TWO:', B, Z
C     .
C     . rest of TWO
C     .

      END
```

I ran the program above just to ensure that BLOCK DATA really works. Here is the output:

```
From ONE:           100         .0000000      -100.0000000
From TWO:mw             .0000000
```

This brings one more point to mind: apparently the uninitialized variables were initialized—even though not all the variables in the DATA statements were included. This is not specified by the FORTRAN 77 standard and is totally dependent on the particular implementation of FORTRAN you are using. In other words, don't depend on COMMON variables being zeroed out for you.

Put It Together

Providing a COMMON example that illustrates the point we're making must necessarily be fairly involved. Examples should also extend the teaching/learning environment and provide you with something to build on in the exercises.

The problem that we will investigate together is that of fitting an equation to a set of data points. Of course, this is "real-world" stuff; you gather some data then want to do something with it—such as integration, differentiation, or extrapolation. This is a curve-fitting program. In the exercises you will see how it's applied.

Curve-fitting isn't automatic. Computers can grind out numbers just fine, but it is the user who must attach some significance to the answers. This problem is a case in point; it will accept data, and "fit it" to four equations. By "fit," I mean it will compute two coefficients that will pass a particular type of curve as close as possible to all of the data points. The four equations assume that the data are non-negative and monotonic. These assumptions are not good generally. For instance, this program will not fit periodic data or data with "bumps" in it. That doesn't make it worthless, but it puts the burden of judgment on you, the end user.

Here is the idea behind the program: For each of the four curves, the input data are first transformed, then a straight line "drawn" through the data in transform space, and finally the resulting coefficients of the straight line are retransformed to normal space. The data fitting is done by minimizing the Y-direction errors—the precise algorithm is left as an exercise. The overall program algorithm follows, and the exact transformations are commented on in the program itself.

a) Prompt/read the data points (main program).
b) Loop through the following steps for each curve candidate:
    i)   Transform the input data points (XFORM).
    ii) Compute least square straight line through the transformed data (SQRFIT).
    iii) Untransform the straight line coefficients (UNXFRM).
    iv) Compute an error measure so that a "best" fit can be determined (ERROR).
    v) Display the results of the fit (main program).
    vi) Loop to step (i) for all four curve types (main program).
c) Loop to step (a) for more data (main program).

One additional roadmap might be useful in understanding this code. I call it a Set/Use matrix; it is a variation of the diagram shown earlier and is better suited to documentation than diagrams. You will see in the matrix that follows that subprogram XFORM "Use"s data in the COMMON block named RAW and "Set"s data in XFRM.

Subprogram	RAW	XFRM	COEFF
MAIN	S/U	S	S
XFORM	U	S	
UNXFRM	S/U		S
SQRFIT		S/U	S
ERROR	U	U	

The definition of "Set" is: One or more variables in the COMMON block is in a READ statement, is on the left of an assignment statement, or is a subprogram output argument. The definition of "Use" is: One or more variables in the COMMON block is in a WRITE statement, is used in an arithmetic or relational expression, or is a subprogram input argument. This sort of information is very useful when you are trying to modify a program.

```
      PROGRAM FITIT
C
C Curve fitting monotone, convex data.
C
C The following curves are fitted through the data. The method is
C to modify the input data (x,y) by a "linearize" operation to
C yield the transform data (x', y'), fit a line through that data to
C compute aslope (BHAT) and intercept (AHAT), then "unlinearize"
C these to coefficients to find the coefficients of the target
C curve. The following table summarizes these operations:
C
C      Target              Data Transform          Coefficient Transform
C      Equation              Equations                  Equations
C      ----------          ---------------         ---------------------
C 1.  y = A+B*x        x' = x       y' = y        A = Ahat      B = Bhat
C 2.  y = A+B/x        x' = 1/x     y' = y        A = Ahat      B = Bhat
C 3.  y = A*e**(B*x)   x' = x       y' = ln(y)    A = e**Ahat B = Bhat
C 4.  y = A*x**B       x' = ln(x)   y' = ln(y)    A = e**Ahat B = Bhat
C
C <> <> <> <> <> <> <> <> <> <> <> <> <> <> <> <>
C Raw COMMON data
C                      MAXRAW is the maximum number of input pairs
      INTEGER MAXRAW
      PARAMETER (MAXRAW=20)
C                      NRAW is the actual number of input pairs
      INTEGER NRAW
C                      XRAW and YRAW are the input data
C                      NEWY is the new Y-values, based on the
C                      fitted curve
      REAL XRAW(MAXRAW), YRAW(MAXRAW), NEWY(MAXRAW)
      COMMON /RAW/ NRAW, XRAW, YRAW, NEWY
C <> <> <> <> <> <> <> <> <> <> <> <> <> <> <> <>
C Transformed COMMON data
C                      MAXFRM is the maximum number of
C                      transformed pairs
C                      It should be the same as MAXRAW
```

```
          INTEGER MAXFRM
          PARAMETER (MAXFRM=20)
C                         NXFRM is the actual number of transformed
C                         pairs. It is the same as NRAW
          INTEGER NXFRM
C                         XXFRM and YXFRM are the transformed data
C                         XNUY is the new Y-values in transform space
          REAL XXFRM(MAXFRM), YXFRM(MAXFRM), XNUY(MAXFRM)
          COMMON /XFRM/ NXFRM, XXFRM, YXFRM, XNUY
C <> <> <> <> <> <> <> <> <> <> <> <> <> <> <> <>
C Coefficients COMMON data
C                         AHAT and BHAT are the intercept and slope
C                         of the best fit straight line through the
C                         transformed data
          REAL AHAT, BHAT
C                         ALPHA and BETA are the coefficients of the
C                         target equation
          REAL ALPHA, BETA
          COMMON /COEFF/ AHAT, BHAT, ALPHA, BETA
C <> <> <> <> <> <> <> <> <> <> <> <> <> <> <> <>
C Local data
C                         TOTFRM is the number of transformations
C                         defined
          INTEGER TOTFRM
          PARAMETER (TOTFRM=4)
C                         NOXFRM is the number of the current
C                         transform
          INTEGER NOXFRM
C                         ACTERR and XFMERR are the error measures
C                         in real space and in transform space
C                         respectively
          REAL ACTERR, XFMERR
C                         CURVE is the display version of the
C                         curve's equation
          CHARACTER*12 CURVE(TOTFRM)

          DATA CURVE(1) / 'y = A + Bx' /
          DATA CURVE(2) / 'y = A + B/x' /
          DATA CURVE(3) / 'y = Ae**Bx' /
          DATA CURVE(4) / 'y = Ax**B' /
```

(continued)

```
C Prompt/read input points
100      CONTINUE
         WRITE (*,*) 'Enter number of data points'
         READ (*,*) NRAW
         WRITE (*,*) 'Enter X-Y pairs'
         READ (*,*) ( XRAW(I), YRAW(I), I=1, NRAW )
         NOXFRM = 1
C
C Display the header for the results
         WRITE (*,*)
         WRITE (*,*)
     +             ' Curve         A        B      Actual   Transform'
C               123456789012123.12345123.12345xx123.12345123.12345
         WRITE (*,*)
     +             '                                    Error    Error'
C
C Transform the data
200      CONTINUE
         CALL XFORM ( NOXFRM )
C
C Compute least square fit coefficients
         CALL SQRFIT
C
C Un-transform the linear coefficients and re-compute Y values
         CALL UNXFRM ( NOXFRM )
C
C Compute the error introduced by the fit
         CALL ERROR ( ACTERR, XFMERR )
C
C Display error and coefficients of the derived fit
         WRITE (*, 10000) CURVE(NOXFRM), ALPHA, BETA, ACTERR, XFMERR
10000    FORMAT ( 1X, A12, 2F9.5, 2X, 2F9.5 )
C
C Loop for next transformation
         IF ( NOXFRM .LT. TOTFRM ) THEN
            NOXFRM = NOXFRM + 1
            GOTO 200
         ENDIF
C
C Loop for next data set
         GOTO 100

         END
```

```
******************************************************
C Transform the raw data according to the curve that
C is being assumed for the data
      SUBROUTINE XFORM ( NOXFRM )

C                              NOXFRM is the transformation number
      INTEGER NOXFRM
C <> <> <> <> <> <> <> <> <> <> <> <> <> <> <> <>
C Raw COMMON data
      INTEGER MAXRAW
      PARAMETER (MAXRAW=20)
      INTEGER NRAW
      REAL XRAW(MAXRAW), YRAW(MAXRAW), NEWY(MAXRAW)
      COMMON /RAW/ NRAW, XRAW, YRAW, NEWY
C <> <> <> <> <> <> <> <> <> <> <> <> <> <> <> <>
C Transformed COMMON data
      INTEGER MAXFRM
      PARAMETER (MAXFRM=20)
      INTEGER NXFRM
      REAL XXFRM(MAXFRM), YXFRM(MAXFRM), XNUY(MAXFRM)
      COMMON /XFRM/ NXFRM, XXFRM, YXFRM, XNUY

      NXFRM = NRAW
C
C CASE on the transformation number
      IF (NOXFRM .EQ. 1) THEN
C    1:
          DO 215 I=1, NXFRM
              XXFRM(I) = XRAW(I)
              YXFRM(I) = YRAW(I)
215       CONTINUE
C    2:
      ELSEIF (NOXFRM .EQ. 2) THEN
          DO 225 I=1, NXFRM
              XXFRM(I) = 1.0 / XRAW(I)
              YXFRM(I) = YRAW(I)
225       CONTINUE
```

(continued)

```
C   3:
          ELSEIF (NOXFRM .EQ. 3) THEN
              DO 235 I=1, NXFRM
                  XXFRM(I) = XRAW(I)
                  YXFRM(I) = LOG (YRAW(I))
235           CONTINUE
C   4:
          ELSEIF (NOXFRM .EQ. 4) THEN
              DO 245 I=1, NXFRM
                  XXFRM(I) = LOG (XRAW(I))
                  YXFRM(I) = LOG (YRAW(I))
245           CONTINUE
C
C   end CASE
          ENDIF

          END


*******************************************************
C Un-transform the coefficients and compute the resulting Y-values
          SUBROUTINE UNXFRM ( NOXFRM )

C                          NOXFRM is the transformation number
          INTEGER NOXFRM
C <> <> <> <> <> <> <> <> <> <> <> <> <> <> <> <>
C Raw COMMON data
          INTEGER MAXRAW
          PARAMETER (MAXRAW=20)
          INTEGER NRAW
          REAL XRAW(MAXRAW), YRAW(MAXRAW), NEWY(MAXRAW)
          COMMON /RAW/ NRAW, XRAW, YRAW, NEWY
C <> <> <> <> <> <> <> <> <> <> <> <> <> <> <> <>
C Coefficients COMMON data
          REAL AHAT, BHAT
          REAL ALPHA, BETA
          COMMON /COEFF/ AHAT, BHAT, ALPHA, BETA
C CASE on transform number
C   1:
          IF (NOXFRM .EQ. 1) THEN
              ALPHA = AHAT
              BETA = BHAT
              DO 315 I=1, NRAW
                  NEWY(I) = ALPHA + BETA * XRAW(I)
315           CONTINUE
```

```
C    2:
          ELSEIF (NOXFRM .EQ. 2) THEN
              ALPHA = AHAT
              BETA = BHAT
              DO 325 I=1, NRAW
                  NEWY(I) = ALPHA + BETA / XRAW(I)
325           CONTINUE
C    3:
          ELSEIF (NOXFRM .EQ. 3) THEN
              ALPHA = EXP (AHAT)
              BETA = BHAT
              DO 335 I=1, NRAW
                  NEWY(I) = ALPHA * EXP ( BETA * XRAW(I) )
335           CONTINUE
C    4:
          ELSEIF (NOXFRM .EQ. 4) THEN
              ALPHA = EXP (AHAT)
              BETA = BHAT
              DO 345 I=1, NRAW
                  NEWY(I) = ALPHA * XRAW(I) ** BETA
345           CONTINUE
C
C    end CASE
          ENDIF

          END

*********************************************************
C Fit a straight line through the transformed data points and
C recompute the transformed Y-values
          SUBROUTINE SQRFIT

C <> <> <> <> <> <> <> <> <> <> <> <> <> <> <> <>
C Transformed COMMON data
          INTEGER MAXFRM
          PARAMETER (MAXFRM=20)
          INTEGER NXFRM
          REAL XXFRM(MAXFRM), YXFRM(MAXFRM), XNUY(MAXFRM)
          COMMON /XFRM/ NXFRM, XXFRM, YXFRM, XNUY
C <> <> <> <> <> <> <> <> <> <> <> <> <> <> <> <>
C Coefficients COMMON data
          REAL AHAT, BHAT
          REAL ALPHA, BETA
          COMMON /COEFF/ AHAT, BHAT, ALPHA, BETA
```

(continued)

```
C <> <> <> <> <> <> <> <> <> <> <> <> <> <> <> <>
C Local data
       DOUBLE PRECISION SUMX, SUMY, SUMX2, SUMXY, DETER

C
C Compute SUM of X's, SUM of Y's, SUM of X*X's and SUM of X*Y's of
C the transformed data
       SUMX = 0.0
       SUMY = 0.0
       SUMX2 = 0.0
       SUMXY = 0.0
       DO 100 I=1, NXFRM
           SUMX = SUMX + DBLE(XXFRM(I))
           SUMY = SUMY + DBLE(YXFRM(I))
           SUMX2 = SUMX2 + DPROD(XXFRM(I), XXFRM(I))
           SUMXY = SUMXY + DPROD(XXFRM(I), YXFRM(I))
100    CONTINUE
C
C Compute coefficients of a straight line: Y = AHAT + BHAT*X, that
C minimizes the Y-distance error
       DETER = ( NXFRM*SUMX2 - SUMX*SUMX )
       AHAT = ( SUMX2*SUMY - SUMX*SUMXY ) / DETER
       BHAT = ( NXFRM*SUMXY - SUMX*SUMY ) / DETER
C
C Re-compute the transformed Y-values based on the straight line
C coefficients that were just found.
       DO 200 I=1, NXFRM
           XNUY(I) = AHAT + BHAT * XXFRM(I)
200    CONTINUE

       END


*****************************************************
C Compute the error factor in the Y-direction of both the
C actual and transformed data
       SUBROUTINE ERROR ( ACTERR, XFMERR )

C                        ACTERR and XFMERR are the error measures
C                        in real space and transform space
C                        respectively
       REAL ACTERR, XFMERR
C <> <> <> <> <> <> <> <> <> <> <> <> <> <> <> <>
```

```
C Raw COMMON data
      INTEGER MAXRAW
      PARAMETER (MAXRAW=20)
      INTEGER NRAW
      REAL XRAW(MAXRAW), YRAW(MAXRAW), NEWY(MAXRAW)
      COMMON /RAW/ NRAW, XRAW, YRAW, NEWY
C <> <> <> <> <> <> <> <> <> <> <> <> <> <> <> <>
C Transformed COMMON data
      INTEGER MAXFRM
      PARAMETER (MAXFRM=20)
      INTEGER NXFRM
      REAL XXFRM(MAXFRM), YXFRM(MAXFRM), XNUY(MAXFRM)
      COMMON /XFRM/ NXFRM, XXFRM, YXFRM, XNUY

      ACTERR = 0.0
      XFMERR = 0.0
      DO 100 I=1, NRAW
          ACTERR = ACTERR + ( NEWY(I) - YRAW(I) ) ** 2
          XFMERR = XFMERR + ( XNUY(I) - YXFRM(I) ) ** 2
100       CONTINUE

      END
```

When FITIT is run, the following dialogue takes place:

```
Enter number of data points
6
Enter X-Y pairs
1 6.88
2 5.03
4 4
8 3.55
15 3.25
30 3.1
```

Curve	A	B	Actual Error	Transform Error
y = A + Bx	5.20347	-.09018	5.42346	5.42346
y = A + B/x	3.01936	3.89563	.01039	.01039
y = Ae**Bx	5.07655	-.02067	5.04388	.20222
y = Ax**B	6.10694	-.22879	.97903	.04129

Pitfalls

The COMMON construct is certainly a very powerful way to share variables, but you must also be aware of its shortcomings, for they can be the source of difficult bugs. Keep this in mind: Each time a COMMON construct is encountered by FORTRAN it is treated independently from all the others. In other words, FORTRAN does *no* cross-checking to ensure that the COMMON construct is the same each time. This means that all sorts of errors can creep into your program via the COMMON declaration:

1. The "var_list" may be ordered differently.
2. The "var_list" may have missing or extra variables in it.
3. The declarations of the variables in the "var_list" may differ. For instance, an array declaration may have different lengths.
4. The "var_list" may have spelling errors in it.

None of the problems above will be noticed by FORTRAN; instead, your program will produce erratic results.

The safest way to work with COMMON is never to trust oneself to enter the COMMON block and its declarations *twice*. Instead:

1. Enter and *group* all the declarations for a COMMON block together in the program, then
2. Use your editor to *copy* the several lines associated with the COMMON declaration to all subprograms using it.
3. In a similar way, make changes only to a single COMMON block, then
4. Copy that modified block, replacing all previous definitions, into all the subprograms.

In the days of IBM cards and batch processing we would keypunch the COMMON declaration, then make several copies of the card deck with the card duplicator and store the copies in a card file cabinet. Each time we started coding a new subprogram, we extracted a duplicated deck and inserted it in the proper place in the subprogram. This saved not only keypunch time but debugging time, because we *knew* that the COMMON declarations were correct.

Apparently there is no "standard" way to handle uninitialized BLOCK COMMON variables; as noted, implementation varies from vendor to vendor. A suggestion you might consider is to initialize them all, just to be on the safe side.

Exercises

1. Enter FITIT into your system and rerun it with the data presented in the "Put It Together" section to be sure you entered it right. Then run the following data through your version of FITIT. You will notice that the output format is not very pretty. This is very much a real-life problem: You will be expected to take a "running" program and put the finishing touches on it. Modify your version of FITIT to clean up the messy program I gave you to work with.

a) 0.5   1.7
     1.0   4.3
     1.3   6.1
     3.0   18.0
     3.4   21.0
     6.0   44.0
     6.25   47.0

b) 9.0   180.0
     1.0   3.29
     3.0   9.0
     2.0   5.44
     6.0   40.1

c) 6.0   19.112
     9.0   28.09
     2.0   7.1
     3.0   9.95

d) 1.5   7.5
     3.0   27.4

2. As FITIT stands in "Put It Together," some of the transformations don't work with zero and negative numbers. You are likely to run into this problem when you leave the academic community—maybe you have already experienced it. In any case, there is a saying that when the programmer announces his or her program is working, it means that half the bugs have been found. Modify FITIT to work even when non-positive numbers are fed into it. This could be done by skipping the offending transformation or by modifying either the data or the transformation to work properly when ill conditioned data are presented to it. Here are some very offensive data sets:

a)	−3	23
	−1	7.7
	1.5	−11.5
	14	−105
b)	−5	−50
	0	−20
	3	−4
	3.5	−1
	3.67	0
c)	−4.5	0.00001
	−3	0.0006
	−1	0.25
	0.5	19
d)	1	1
	1	9
	1	−3.7

3. Other transformations could be added to FITIT to make it even more general—and therefore useful. Modify FITIT to test for and display the results of one or more of the following curves:

Target Equations	Data Transform Equations	Coefficient Transform Equations	
a) $y = A + B*\ln(x)$	$x' = \ln(x)$ $y' = y$	$A = Ahat$	$B = Bhat$
b) $y = A/(B+x)$	$x' = x*y$ $y' = y$	$A = -Ahat/Bhat$	$B = -1/Bhat$
c) $y = A*x/(B+x)$	$x' = y/x$ $y' = y$	$A = Ahat$	$B = -Bhat$

4. Write a paper describing the straight-line fitting method used. A good reference is Maron.

5. The method above minimizes the error in the Y-direction only. Another technique is to minimize the error perpendicular to the line instead. Derive the necessary equations.

6. Take the results of the previous problem (Exercise 5) and modify FITIT. Then compare the results of the two methods. Test data can be generated by starting with an equation in one of the known forms, picking some coefficients, plugging some values of X into that equation and then adding or subtracting a small value to the resulting value of Y.

7. This is an exercise to put together various software packages. Combine FITIT with one of the plot routines you developed in Chapter 7. Plot the raw data and the data resulting from the four curve fits for one of the data sets in Exercise 1. If you haven't already done so, this is a good opportunity to "subprogramize" your plot routine to make it nice and portable. You will find lots of uses for it if it is easy to use. Now that you have a better grasp of COMMON, you may want to modify your plot routine to shorten the calling sequences.

8. This is an exercise in putting together several software packages to yield still another tool. You are to combine FITIT with one of the integration packages you developed in Chapter 5. Here is the algorithm I would like your program to perform:
a) Prompt/read data points.
b) Produce coefficients for the four curves and display the results.
c) By examining the error values, select the best curve.
d) Prompt/read the integration limits.
e) Integrate the best fit curve over the requested limits.
f) Display the result.
g) Loop to step (a).
To test this program, you can either use data from Exercise 1 or derive your own as in Exercise 6 to derive the input data. Then, knowing which equation your program selected, and its coefficients, you can analytically integrate the formula to compare to your program's solution.

10

FILES

In general terms, files allow you to extend memory resident data structures—which are limited by the amount of memory available—to external memory, usually disk, thus greatly increasing the capacity of that data structure. Again, in general terms, READ is the construct that allows the programmer to transfer data from the external medium to memory, and WRITE is the construct used to transfer data to the external medium. Because of the various ways that operating systems actually implement files there are many variations to this general model. We will see later in this chapter that FORTRAN has variants too—in particular, the concept of "internal files."

One's first exposure to files can be very perplexing, and even seasoned programmers schedule extra time when they are designing a program that includes files. Perhaps a brief introduction to terms can help you through the early anxiety stages. A "file" of data is made up of zero or more "records." A "record" is the smallest unit of data that can be controlled in the file; usually it contains several variables. A data file has two primary structural attributes:

1. Access method (how to reach data in the file): either "sequential" or "direct."
2. Record type (method of data storage): either "formatted" or "unformatted."

Standard FORTRAN 77 combines these attributes all four ways: sequential-formatted, sequential-unformatted, direct-formatted, and direct-unformatted. The mechanics and use of each of these combinations will be discussed in the following three sections. Then, in Section 10.4, their features will be compared.

An overview of possibilities may help to show you where we are heading. A programmer may perform several operations on files. Consider the following scenarios as typical, but not as a complete list of everything that might be done. Usually, when creating a file, the sequence used is:

1. Identify the file for the operating system, using OPEN.
2. Repeatedly WRITE records into the file.
3. When the file is completely written, signal its end with ENDFILE.
4. Tell the operating system to CLOSE the file, thus making it unavailable for further operations.

Another "typical" operation is using an existing file strictly as a dictionary—that is, using it without performing any modifications to the data. In that case, the following sequence may take place:

1. Identify the desired file using OPEN.
2. Repeatedly READ records from the file, performing some function using the data.
3. Tell the operating system to CLOSE the file.

Finally, another possible operation the programmer may want to perform is to modify, update, and change some of the records in an existing file, thus creating a new one. In this case, a typical sequence might be:

1. OPEN the old file.

2. OPEN the new file.
3. Create the new file using the old file and other data by any combination of the following:
   a) Select and READ an old record, then modify it and WRITE it on the new file, or
   b) READ an old record and WRITE it to the new file, unmodified, or
   c) Insert a new record using WRITE into the new file.
4. CLOSE the old file.
5. ENDFILE the new file.
6. CLOSE the new file.

One final word of caution: The presentation of the READ/WRITE construction in this chapter is fundamental. Most FORTRANs have extended the standard; you will probably find other variations and additions in the FORTRAN version you are working with. To further confuse the picture, the various FORTRANs also have slight differences in nomenclature. For instructional purposes, we have cut through these inconsistencies to describe the uniform properties and uses of files. The text and exercises in this chapter are concerned only with standard structures. However, as you will see, these primary constructions are very powerful and most applications can be supported with them.

10.1 RECORD TYPES: FORMATTED AND UNFORMATTED

When we speak of formatted and unformatted records or files, we mean whether the records contain numbers in ASCII format or in internal machine format. While this is a formal definition, it is not a very useful one. Let us point out the implications.

In a formatted record, the numbers are in ASCII format, so they must be *converted* into INTEGER or REAL format before any mathematical operations can take place. Up to this point in the text, you have been working primarily with formatted records—whether or not you specified a FORMAT statement. That is, formatted READ/WRITE statements had one of the following forms:

```
READ (*,*)
READ (*,label)
WRITE (*,*)
WRITE (*,label)
```

More precisely, when performing terminal I/O, you are doing sequential, formatted operations.

On the other hand, operations on unformatted records need no FORMAT specification. This form of the READ/WRITE is new—it has not been discussed yet. It is very simple and will be defined more fully in the sections that follow. For now however, look at the basic form:

```
READ (unit)
WRITE (unit)
```

Here is another implication to the formatted/unformatted definition: Only the formatted file can be displayed/listed/printed by system utilities like PRINT and TYPE. In order to display the contents of an unformatted file, you must write a routine that contains the necessary READ/WRITE/FORMAT statements. A further result of this definition is that formatted files can be created with your editor, but unformatted files must be created with a user-written routine.

Since we have made a good case in favor of formatted files, you would logically expect that the unformatted file must have some advantages over formatted files. And there are good reasons. Since an unformatted file does not have to pass through the FORMAT interpreter, unformatted data can be processed faster, allowing your program to run faster. The FORMAT interpreter is a very sophisticated and very involved chunk of software; to be executed, even the most trivial of FORMATs requires many hundreds of instructions. In avoiding the FORMAT interpreter, your program can be running more of the time. You may enjoy an exercise at the end of this chapter that calls for you to investigate the FORMAT interpreter timing overhead.

Another reason for unformatted files is to maintain real number precision. A REAL requires only 4 bytes in an unformatted file, but would require 13 bytes in a formatted file (s.xxxxxxxEsyy), and there is the possibility that round-off errors will creep into this conversion. So, unformatted files offer both size and accuracy advantages for REAL numbers.

One final point in favor of unformatted files is that, by clever designing, you can also save space in your file for other data types. This topic will be continued in Appendix E, under the DIRECT-OPEN discussion. Briefly, however, short records are superior to long records for two reasons.

1. Short records are transferred faster between the computer and the storage medium. It takes a constant amount of time to move each byte, so, logically, the fewer number of bytes moved, the faster the

movement of data will take place. In a very simplistic view of how a computer works, the computer can either move data or execute instructions so, if it is moving data, is can't be executing your program. The better programs minimize data transfer time by minimizing the amount of data transferred.

2. Shorter records mean either that the files are smaller, or that more data can be stored in the same amount of room. Even though external media—tapes and disks—continue to have greater and greater capacities, there are limits. Capacities are finite.

10.2 SEQUENTIAL FILES

Sequential files have a long history and have been with FORTRAN from the very beginning. Tapes (and occasionally drums) were the first external storage media. Disks were not available until the mid-1960s. Early FOR-TRAN programs dealt only with records. The concept of a named file was introduced with disks. A programmer would consider a whole tape a sin-gle file, and usually only one file was on a tape. When the programmer's tape was "hung" on a tape drive, the operator or the programmer would "dial" the tape drive to the appropriate number—from 0 to 7. This meant that a statement like "REWIND 7" actually caused the tape on the unit 7 tape drive to be repositioned to the beginning—a truly spectacular event.

In the early 1960s, very few data centers allowed on-line input or out-put because these are relatively slow processes. On-line computers were not even connected to card readers or line printers. All input and output was via tapes. So, in order to feed data to and get listings from the on-line computer, a smaller, slower, less expensive off-line computer was dedicated to the chore of "card-to-tape" and "tape-to-print" operations. Then, to accommodate this scheme, computer centers had a standard con-vention for tape drive numbers. For instance, on IBM systems, zero was the system tape; it contained the FORTRAN compiler. Five was the sys-tem input tape; it contained the programmer's FORTRAN deck and test data. And six was the system output tape—the listing tape. So program-mers had five units (1 through 4 and 7) at their disposal. You may see these conventions carried into some of today's systems, and now you will un-derstand the historical reasoning.

Here is a list of sequential file characteristics and a discussion of each of them. When you read the following, it may help you to think of a cassette tape:

1. Records in a sequential file must be accessed (read or written) one after another, in the order in which the file was created. This is analogous to the cuts of music on a cassette tape. To get to the fourth cut, you must either play or fast-forward past the first three.
2. Records in a sequential file may have different lengths. This, too, is like a cut on a cassette tape.
3. Records in a sequential file may *not* be erased. Here the analogy is starting to break down; you can use some tricks to erase cassette cuts.
4. Records in a sequential file may not be overwritten. You can go back to the cassette analogy to see the problem. When you re-record a cut in the middle of a cassette, you get into various problems, depending on the length of the new cut compared to the length of the old one. Rather than solving these problems, FORTRAN 77 simply erases all records in the file that follow the overwritten one, effectively destroying the remainder of the file.
5. Even though sequential files were originally developed for tapes, FORTRAN 77 now supports both tape and disk sequential files.
6. The record operations supported by FORTRAN 77 are:
   - BACKSPACE   reposition to previous record.
   - READ   transfer a file from external medium (disk or tape) to memory.
   - WRITE   transfer a file from memory to external medium (disk or tape).
7. The file operations supported by FORTRAN 77 are:
   - CLOSE   opposite of OPEN.
   - ENDFILE   WRITE a special record that signals the end of the file.
   - INQUIRE   examine properties of a file.
   - OPEN   prepare a file for access.
   - REWIND   position file to beginning.

A detailed description of each of these operations can be found in Appendix E.

10.3 DIRECT FILES

The list of direct-file characteristics that follows is in the same order as the list in the sequential file discussion; you may care to compare the two.

Direct files were developed primarily to take advantage of the capabilities provided by disk hardware. The analogous model to keep in mind, in the following presentation, is the phonograph record.

1. Records in a direct file may be accessed either randomly by record number or sequentially, as described in the previous section. This is analogous to a phono record: You can play any cut by placing the arm in the desired location, or you can play the whole side in the order of the cuts.

2. All record sizes in a direct file *must* be the same length. This restriction results from the way the data are stored and retrieved from the disk. Our phonograph record analogy falls apart in this respect, since the length of music cuts can vary. This restriction lays additional burdens on the programmer, as will be discussed shortly.

3. Records in a direct file may not be erased. The implications of this restriction go further than one might expect. Suppose you OPEN a "new" file and immediately WRITE record number 10. There is an unexpected side effect: You have also created records 1 through 9 and *all have random data in them*. You, the programmer, must keep track of the fact that only record number 10 has valid data, and that the other 9 are garbage. But, don't stop here; look at the next characteristic!

4. Records in a direct file *may be* overwritten. This is a very nice feature and is exactly opposite to the philosophy of sequential records. This capability allows programmers to WRITE direct records in any order and also allows one or more records to be replaced without copying the entire file.

5. Direct files may not reside on tape, but only on disks. This restriction is primarily a mechanical one due to capability 4. It is difficult to rewrite a record on tape without the risk of running into the following record. In any case, because tape is a sequential medium by its very nature, direct access doesn't make much logical sense.

6. The record operations supported by FORTRAN 77 are:
   - READ   transfer a file from external medium (disk) to memory.
   - WRITE   transfer a file from memory to external medium (disk).

7. The file operations supported by FORTRAN 77 are:
   - CLOSE   opposite of OPEN.
   - INQUIRE   examine properties of a file.
   - OPEN   prepare a file for access.

A detailed description of each of these operations can be found in Appendix E.

10.4   SEQUENTIAL-DIRECT COMPARISON

It is important for you to consider the pros and cons of the four file structures that have been discussed, because:

1. The performance of your program is greatly affected by the file type—"unformatted/direct" is usually fastest.
2. The amount of software that has to be created is affected by the file type—"formatted/sequential" usually requires the least.
3. The amount of external medium space you have available may be an issue—there is no clear cut choice.
4. The identity of the end user of that data structure is important— "unformatted" is used for program-to-program interfacing and "formatted" is used to interface to the real world.

The discussion that follows includes some general comments about some applications, but there are numerous exceptions; so, in the end, you will have to make the design decision.

SEQUENTIAL/FORMATTED

This structure can be used in the place of terminal input/output operations. This is easiest to do because you are already used to it; it is coded just like terminal I/O. Sequential/Formatted has a very real advantage in that such files can be created with the system editor and displayed or printed with system utilities. However, this is the slowest of the four methods.

SEQUENTIAL/UNFORMATTED

This structure is useful for trivial program-to-program interfacing—for instance, to pass a very large array from one program to another. This structure doesn't normally "talk" to the outside world very well: You cannot create the file with the system editor and you cannot display the file on your terminal or on a printer; therefore, auxiliary software must be created to do display or listing. There is one exception, however: If the file contains *no* numbers—that is, if it is all CHARACTER data—there is no need for FORMAT statements. In this case the system editor may be used and the file may be listed or displayed.

DIRECT/FORMATTED

This structure probably will not appear very often. If you need a direct file, you might as well go to the slight additional trouble of making it "unformatted." The exception to this statement could be when the records contain many small integers; you may save space by using a "formatted" structure. A more detailed discussion of size is found in Appendix E, under DIRECT-OPEN.

DIRECT/UNFORMATTED

This structure is to be expected in most database applications: it is the fastest structure to access randomly. By "database application" we mean those applications that primarily READ data in a random manner—any application that uses a data file as a "dictionary." However, whenever you design such an application, you must also design a companion application to "maintain" the database—to add, change, delete, and list records in the database—because the system doesn't usually support maintenance of direct files. VAX/VMS® is an exception to this statement; Digital Equipment Corporation has included many fine utilities to support direct and other nonstandard file types.

A subtle point concerning direct files should be noted: Most applications don't lead to direct-file record numbers easily. This statement is not very obvious, so study the following example. Suppose you wanted to design an "address book" application program. After thinking about the problem some, you would probably realize the following.

1. Memory is too small for the amount of data you would like to enter, and so the data must be stored on disk.
2. The most common operation you would like to perform would be to input a friend's last name and have your program display a phone number and an address.
3. It will be necessary to add, change, and delete entries in your address book database from time to time.

The obvious disk data structure is to store the name, address, and phone number for each person, sequentially, in its own record; this will accommodate observation (1). However, the order of these records is essential to observation (2). This may not be so obvious, but let's consider possible search algorithms.

First, let's use a sequential search on an unordered file.

1. Start at record number 1 in the file.
2. Read a record. If at the end of the file, stop.
3. Examine the record. If that is the desired record, display it and stop.
4. If not the desired record, return to step (2).

This method is easy to code, but could be time-consuming to execute. For instance, if you make a spelling error, the entire file is searched anyway. Even in the average case, half of the file must be searched.

Second, consider searching an alphabetically ordered file, sequentially.

1. Start at record number 1 in the file.
2. Read a record. If at the end of the file, stop.
3. Examine the record. If it is the desired record, display it and stop. If it is too far in the file, stop.
4. If it's not the desired record, return to step (2).

The change in this method is in step (3); in case of failure, the program doesn't necessarily continue searching until the very end of the file. Even so, these algorithms suffer in other aspects:

1. Each examination requires a disk access, and this is time-consuming.
2. Any add, change, or delete to the file requires that the file be entirely rewritten, due to the nature of sequential files.

The basic characteristic of this address book application is that the data are to be accessed randomly, not sequentially. Random access implies that direct files rather than sequential files should be used, and the question is, "How?" Answer: "By separating the name from the rest of the record." Here are the details:

1. Create a second data structure to contain only the names. The goal of this move is to keep the "key" data—the data to be searched—in memory, thus greatly reducing the search time.

2. Create a third data structure to contain the record number of the complete record. These data are kept in memory, too.

Here is a new algorithm using these two data structures. As you read the algorithm, refer to the following diagram, too:

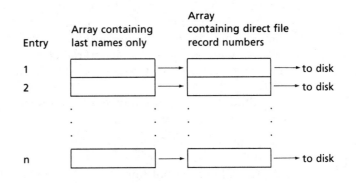

1. Start at entry number 1 in the array of last names.
2. Examine the entry. If it is the desired record:
   a) Fetch the record number from the corresponding entry in the record number array.
   b) Using that record number, direct access the file to read the complete data record.
   c) Display that record and stop.
3. If it is not the desired name, go to the next entry, if any, and return to step (2). If at the end of the name array, stop.

You will probably notice that other search algorithms could be used on the "name array"; if it is ordered, a binary search could be used to speed up the search algorithm even more. In the "Put It Together" section, a third search method, called "hashing," is introduced.

A second benefit of direct-file data storage is that modification to the database is much easier because of the nature of direct files. For instance, a change is trivial: Read the record, change it, and rewrite that single record. Add and delete, although not as trivial, are much easier and faster to perform than if your design is based on a sequential file.

In summary, direct-file applications are generally accompanied by small arrays, which contain only the "key" data. These arrays reside in memory while the complete file resides on disk. Searching is performed quickly on the memory-resident "key" array, and only a single disk access is made to disk to retrieve the complete data record.

10.5 INTERNAL FILES

Internal "files" are yet another class of files FORTRAN 77 provides. In this case, an external medium is not involved. Instead, the data are merely moved from one place in memory to another. Through internal files, the programmer is allowed to convert data through the FORMAT interpreter. The characteristics of the internal file are single record and FORMAT-TED.

Although there are many reasons for needing this form, the three most common reasons are:

1. To build an output line from right to left, rather than from left to right.
2. To build an output line at several points within your program.
3. To reexamine the input line.

The way in which a programmer creates and uses an internal file is first to define a CHARACTER variable to hold the record; it will be called "buffer" in the remaining discussion. Only two operations can be performed on internal files, READ and WRITE. The syntax for these two special cases is:

READ (buffer, format) I/O_list

WRITE (buffer, format) I/O_list

---

where the following definitions apply:

- "buffer" is a CHARACTER string variable large enough to receive the data.
- "format" is exactly as defined in Chapter 6.
- "I/O_list" is exactly as defined in Chapter 6.

A READ converts and transfers data from the buffer to the variables in the I/O list, under FORMAT control. The WRITE converts and transfers data from the variables in the I/O list to the buffer.

Put It Together

This chapter contains the longest examples in the book. This is necessary to show adequately how files are used. This problem also gives me an opportunity to introduce you to "hashing," a very useful technique for "random" file indexing.

This program is a "weight watcher"; it computes the number of calories consumed in a meal. Even though this is not a difficult concept, the data structure is not trivial, and we will take the time to detail that structure for you. But first, let's examine the overall algorithm of the program:

a) Initialize.
b) Prompt/read for a food.
c) Look up that food in the data base—the calorie table.
d) If that food is not in the data base, display error message and go to step (b).
e) Otherwise, display the calorie count for that food.
f) Prompt/read the quantity of food consumed at the meal.
g) Compute/accumulate and display calories for that food.
h) If more foods were in the meal, loop to step (b).
i) Otherwise, display accumulated calories for meal and stop.

CALORY contains four subprograms. A detailed description of the file structure and a description of each of the subprograms—how each implements and supports this structure—is found in the next several paragraphs. The files are arranged somewhat like the "address book" application.

The file structure used for this application is of the "two-level" type. The first level is a one-record sequential file called CAL_DIR.DAT, which is a "directory" used to access the complete data. The second level is a multirecord direct file which contains the complete data and is called CAL_DAT.DAT. Both files are "unformatted," and the specific file attributes are indicated in the file initialization routine, GETDB. GETDB OPENs and copies the directory into memory, then CLOSEs it, since it uses the copy for the remainder of the program. The data file is simply OPENed for later use by GETREC.

As described earlier, the advantage of a two-level structure is this: The directory level is memory resident, thus permitting very fast searches. The data level normally is too large to fit in memory, and searching it on disk would be very time consuming. Instead, the directory is searched, and it, in turn, "points" to the data. A second advantage is that only the relatively small directory is "full," and the data file is only as big as necessary to hold the current data, thus conserving disk space.

The directory is an array of INTEGER*2 called DIR. The nth entry in DIR contains a record number used to read the data file, CAL_DAT.DAT. The technique for computing "n" is called "hashing," and the computation is made in HASHIT. A food is "hashed" by making a fancy summation using each of the characters of the food. This summation is statistically unique and is used to index into DIR to find the data record number corresponding to the food.

The data record is read by GETREC. It is possible that the hash operation failed to generate a unique number—this is called a "collision"—so GETREC must check to see if the record it read was the record it wanted. If not, it uses sequential entries in DIR—that is, DIR(n + 1), DIR(n + 2), etc.—to search for the desired record, until either the record is found or DIR indicates the record is not available. A value of zero in DIR(k) indicates that the search is to stop and that there is no corresponding data. There is never a record number zero in a direct file.

Once the desired "food" record is located, the accompanying caloric units and caloric value are used to complete the dialog needed to compute the actual consumed calorie count.

The function of FODITM is to suppress trailing blanks and to make all the input uniform using upper case. Unfortunately, I had to play a "trick" on FORTRAN in this routine. I had to use the same trick in HASHIT. In those two routines, I referred to FOOD as a 20-element array so I could address individual characters. Elsewhere, I refer to FOOD as a 20-character string, so I can do string comparisons. Notice that FORTRAN neither cares nor comments on this fact! The responsibility for this type-switch is entirely mine.

It is important to know that this program really runs and is not simply an exercise that I invented. Yes, it does run, under MS-DOS®; therefore, there are MS-FORTRAN® language-specific items included. The one I want to point out is found in the FORMAT for 10001; the backslash means "do not return the carriage." Notice how it works in the sample dialog. The file names themselves are another system-specific point—the underbar is permitted by MS-DOS®.

```
          PROGRAM CALORY
C
C Compute calories for a meal
C                    FOOD is the name of the food
          CHARACTER*20 FOOD
C                    STATUS is result of dictionary search. 0=OK
          INTEGER STATUS
C                    UNAME is the unit descriptor
          CHARACTER*10 UNAME
C                    UVALUE is the multiplier corresp to UNAME
          INTEGER*2 UVALUE
C                    UNUMB is the unit number the user selected
          INTEGER UNUMB
C                    UNITS is the number of units in the serving
          REAL UNITS
C                    UCALS is the number of calories in that
C                    serving
          INTEGER UCALS
C                    TOTCAL is the total calories for the meal
          INTEGER TOTCAL
C                    RESPON is user [Y/N] response
          CHARACTER RESPON
C
C 1. Initialize
          CALL GETDB(STATUS)
          IF (STATUS .NE. 0) THEN
              GOTO 999
          ENDIF
          TOTCAL = 0
C 2. Prompt/read the next food
100       CONTINUE
          WRITE (*, 10001) '0Enter a food: '
10001     FORMAT (A, \)
          READ (*, 10002) FOOD
10002     FORMAT (A)
C 3. Look up the food in the dictionary
C    a. Validate FOOD, count and convert to upper case
          CALL FODITM (FOOD, FODLEN)
C    b. Create the hash code of FOOD
          CALL HASHIT (FOOD, FODLEN, HASHCD)
C    c. Get the entry out of the database
          CALL GETREC ( FOOD, HASHCD, UNAME, UVALUE, STATUS )
```

(continued)

```
C 4. If FOOD not found, display error msg and go to step 2
        IF (STATUS .NE. 0) THEN
            WRITE (*,*) 'I don''t know anything about ', FOOD
            GOTO 100
        ENDIF
C 5. If FOOD found in dictionary, display the units and value.
        WRITE (*, 10013) UNAME, ' have', UVALUE, ' calories each.'
10013   FORMAT (1X, 2A, I4, A)
C 6. Prompt/read the amount eaten in the meal
        WRITE (*, 10003) ' How many ', UNAME, ' did you eat? '
10003   FORMAT (3A, \)
        READ (*,*) UNITS
C 7. Compute and display calories for food - accumulate meal total
        UCALS = UNITS * UVALUE
        WRITE (*,*) 'That is', UCALS, ' calories.'
        TOTCAL = TOTCAL + UCALS
C 8. Prompt/read if any more foods
        WRITE (*,10001) ' Any other foods in your meal {Y/N}? '
        READ (*, 10002) RESPON
C 9. If more foods, go to step 2
        IF ( RESPON .EQ. 'Y' .OR. RESPON .EQ. 'y' ) THEN
            GOTO 100
        ENDIF
C 10. If no more foods, display calories for meal
C
        WRITE (*,*) '--------------------------->>'
        WRITE (*,*) 'Total calories for meal was', TOTCAL


999     CONTINUE
        END
*******************************************************************
C Initialize Data Base Files
        SUBROUTINE GETDB ( IOS )


C                       IOS {0} is 0 if OPEN succeeded
        INTEGER IOS


C <> <> <> <> <> <> <> <> <> <> <> <> <> <> <> <>
C DIRectory COMMON
C                       DIR contains the record number into
C                       CAL_DAT
        INTEGER*2 DIR (997)
```

```
C                         LSTREC  is  the  last  record  written  in
C                         CAL_DAT
      INTEGER*2 LSTREC
C
      COMMON /DIRECT/ DIR, LSTREC

C 1. OPEN the detailed data - CAL_DAT.DAT
         OPEN (1, FILE='CAL_DAT.DAT', STATUS='OLD',
     +      ACCESS='DIRECT',   FORM='UNFORMATTED',    IOSTAT=IOS,
     +      RECL=32)
C 2. If OPEN is unsuccessful, bale out!
         IF (IOS .NE. 0) THEN
            WRITE (*,*) 'Oops - the Calorie data base is gone!'
C 3. Else, OPEN the directory - CAL_DIR.DAT
         ELSE
            OPEN (2, FILE='CAL_DIR.DAT', STATUS='OLD',
     +         ACCESS='SEQUENTIAL',   FORM='UNFORMATTED',
     +         IOSTAT=IOS)
C 4.     If OPEN is unsuccessful, bale out
            IF (IOS .NE. 0) THEN
               WRITE (*,*)
     +            'Oops - the Calorie directory is gone!'
C 5.        If OPEN successful, read it in and CLOSE it.
            ELSE
               READ (2) (DIR(I), I=1, 997), LSTREC
               CLOSE (2)
            ENDIF
         ENDIF

      END
*******************************************************
C FOOD ITEM - Validate, count and convert to upper case, a food
C item.
C       1. No test for non-alpha characters.
C       2. 20 characters is the maximum examined by this routine

      SUBROUTINE FODITM (FOOD, FODLEN)

C                         FOOD {I/O} is the food item to be processed
      CHARACTER FOOD(20)
C                         FODLEN {0} is the length of FOOD
      INTEGER FODLEN
      I = 20
```

(continued)

```
C Loop - starting on right, move left to rightmost non-blank
100      CONTINUE
         IF (FOOD(I) .EQ. ' ') THEN
             I = I-1
             GOTO 100
         ENDIF
C Save FOOD length in FODLEN. Loop thru FOOD to make all upper
C case.
         FODLEN = I
         I = 1
200      CONTINUE
         IF (I .LE. FODLEN) THEN
C Convert to UC.  97 is an 'a'. Subtracting 32 from LC makes it UC
             IF ( ICHAR(FOOD(I)) .GE. 97 ) THEN
                 FOOD(I) = CHAR ( ICHAR(FOOD(I)) - 32 )
             ENDIF
             I = I+1
             GOTO 200
         ENDIF

         END
*******************************************************
C HASH the FOOD - isn't that cute?
C 1. Multiply each character by a unique prime and make a SUM.
C 2. Take center of the SUM and form a number between 1 and 997

         SUBROUTINE HASHIT (FOOD, FODLEN, HASHCD)

C                      FOOD {I} is the food item to be processed
         CHARACTER FOOD(20)
C                      FODLEN {I} is the length of FOOD
         INTEGER FODLEN
C                      HASHCD {O} ranges from 1 ... 997
         INTEGER HASHCD
C                      PRIMES are the prime multipliers for the
C                      hash
         INTEGER PRIMES(20)
C                      SUM is the HASHCD accumulator
         INTEGER SUM

         DATA PRIMES /953, 859, 719, 661, 577, 409, 359, 269, 191,
     +               79, 907, 821, 757, 601, 499, 379, 311, 227,
     +               103,  31/
```

```
C Sum each character times its prime
        SUM = 0
        DO 100 I = 1, FODLEN
            SUM = SUM + PRIMES(I) * ICHAR(FOOD(I))
100     CONTINUE
C Take some numbers out of the middle of SUM and
C convert to proper range: 1 ... 997
        HASHCD = MOD (SUM/3, 997) + 1

        END
*********************************************************
C GETREC will get an entry from the DB.

        SUBROUTINE GETREC ( FOOD, HASHCD, UNAME, UVALUE, RESCOD )

C                       FOOD {I} is the food item to be processed
        CHARACTER*20 FOOD
C                       HASHCD {I} ranges from 1 ... 997
        INTEGER HASHCD
C                       UNAME {O} is the calorie unit's name
        CHARACTER*10 UNAME
C                       UVALUE {O} is the calories per unit
        INTEGER*2 UVALUE
C                       RESCOD {O} is the result code. O=GET
C                       succeeded
        INTEGER RESCOD
C <> <> <> <> <> <> <> <> <> <> <> <> <> <> <> <>
C DIRectory COMMON
C                       DIR contains the record number into CAL_DAT
        INTEGER*2 DIR (997)
C                       LSTREC is the last record written in
C                       CAL_DAT
        INTEGER*2 LSTREC
C
        COMMON /DIRECT/ DIR, LSTREC

C                       DBFOOD is FOOD in the database
        CHARACTER*20 DBFOOD

C 1. Initialize loop at HASHCD
        I = HASHCD
C 2. Loop to examine DIR entry, starting at HASHCD
```

(continued)

```
100       CONTINUE
C    A. If DIR entry not empty, READ that record into DBFOOD
          IF ( DIR(I) .NE. 0 ) THEN
              READ (1, REC=DIR(I)) DBFOOD, UNAME, UVALUE
C       a. If record matches FOOD, set RESCOD to ''success''
C          return
             IF (DBFOOD .EQ. FOOD) THEN
                 RESCOD = 0
                 RETURN
C       b. If record doesn't match, increment to next entry -
C          MOD 997 and try again - unless entire directory
C          searched.
             ELSE
                 I = I + 1
                 IF ( I .GT. 997 ) THEN
                     I = 1
                 ENDIF
             ENDIF
C       c. If looped back to HASHCD - the starting point - then
C          set RESCOD is "unpresent"
             IF ( I .EQ. HASHCD ) THEN
                 RESCOD = -1
C       d. Otherwise, loop to examine next directory entry.
             ELSE
                 GOTO 100
             ENDIF
C    B. If DIR entry empty, RESCOD is "unpresent"
          ELSE
              RESCOD = -1
          ENDIF

          END
```

A sample execution, using an existing database, might look like this:

```
Enter a food: coffee
cups        have    7 calories each.
How many cups       did you eat? 2
That is             14 calories.
Any other foods in your meal {Y/N}? y

Enter a food: cantaloupe
QUARTER     have   47 calories each.
How may QUARTER     did you eat? 1
```

```
That is                47 calories.
Any other foods in your meal {Y/N}? y

Enter a food: egg
each           have 79 calories each.
How many each         did you eat? 3
That is               237 calories.
Any other foods in your meal {Y/N}? y

Enter a food: butter
TABLESPOON have 102 calories each.
How many TABLESPOON did you eat? 1
That is               102 calories.
Any other foods in your meal {Y/N}? n
------------------------->>
Total calories for meal was              400
```

Testing CALORY is very straightforward if the hashing algorithm (HASHIT) and the data location (GETREC) algorithms both work correctly—this is left as an exercise. To test the remaining code, the programmer must:

1. Make sure that missing files will cause the indicated messages to be displayed.

2. Test to see that the proper caloric value is passed back to the main routine and that the multiplication takes place properly.

3. Test to see that the accumulation is done correctly.

4. Test to see that missing database items are detected.

5. Test to see that FODITM correctly counts and changes FOOD to upper case.

6. Test the dialog generally—to make sure that the right message comes out at the right time, and so forth.

Pitfalls

Since there are so many ways to do things wrong, good programming practice always makes use of the IOSTAT and ERR options in all possible statements—OPEN, READ, etc. Other common errors are:

1. Wrong length specified for a direct file: found at read/write time, not at open time where it is actually specified.

2. Creating a direct file with one length and then trying to access it with another; the use of INQUIRE is a good way to interface READ and WRITE properly.

3. Variables in the direct READ list different from those in the corresponding WRITE list; in sequential files, this would cause problems too, but the system would probably locate the error for you.

4. Carriage Control needed *only* on printer files. If present in other files, you must skip over it when reading.

5. You can only add to the end of a sequential file; if you insert records anywhere in the middle of the file, the rest of the file will be lost.

6. Insufficient buffer length for internal file.

7. The order of the arguments is wrong.

8. Incorrectly spelled keyword or string constant.

9. Missing statement "label" in ERR=label or END=label.

10. Tic marks missing on strings; for instance, 'KEEP'.

Exercises

Most systems provide easy ways for the user to specify the use of input and output *files* rather than keyboard and screen. For instance, VMS® provides the DEFINE command and MS-DOS® provides the " < " and " > " convention. It is still a good thing to get to know sequential file processing, so the first three problems are designed to give you an opportunity to practice.

1. Rework PLOTTER (Chapter 6) to receive data from a sequential file instead of from the keyboard.

2. Rework FITIT (Chapter 9) to receive data from a sequential file instead of from the keyboard.

3. Provide a friendly interface to either PLOTTER or FITIT to give the user the option of keyboard or file input at run time.

4. Develop a way to compare the timing of "formatted" and "unformatted" I/O that operates only on CHARACTER variables. Consider the following algorithm:
   a) OPEN two sequential files, one FORMATTED and one UNFORMATTED.
   b) Turn on your "system timer."
   c) WRITE 10 records, each containing a CHARACTER∗10 variable into the FORMATTED file.
   d) Turn off the timer and display the elapsed time.
   e) Repeat steps (b) through (d) for the UNFORMATTED file.
   f) If the time difference between the two methods is very small, try WRITEing 100 records instead, or put more variables in a record.
   g) Vary record length and number of records and repeat the above for various values of these parameters.
   h) Prepare a presentation of your results. Graphs are appropriate and so is a narrative description of your conclusions.

5. Develop a way to compare the timing of "formatted" and "unformatted" I/O that operates only on INTEGER variables. See Exercise 4 for suggested algorithm.

6. Develop a way to compare the timing of "formatted" and "unformatted" I/O that operates only on REAL variables. See Exercise 4 for suggested algorithm.

7. This problem is designed to show how well the hashing algorithm in CALORY does. Build a MAIN program that just dialogs with the keyboard, call HASHIT and FODITM and display the results of the hashing. Then, input a number of names to look for a collision; it should occur only very rarely. Here is the algorithm I have in mind:
   a) Prompt/read a word.
   b) FODITM it, then HASHIT.
   c) Display the HASHCD.
   d) Loop to (a).

You are looking for two different words hashing to the same value. Since it may be somewhat difficult to remember all your previous guesses, the program could keep track for you. Use a CHARACTER*20 array to store (at HASHCD) the word you are testing. If a word is already there, you have a collision, and you know the two words that collided.

You cannot do much with CALORY without the database generation routine. I have included a partial listing of my database manager. The remaining problems in this chapter are related to CALORY and to the following code. You need to enter CALDBM and build yourself a database if you expect to do any work on CALORY.

```
        PROGRAM CALDBM
C
C Calorie Data Base Management ...
C
C
C                       RESPON is user response
        CHARACTER RESPON

C I.   Initialize
        CALL GETDB
C II. Prompt/read A/C/D/L
100     CONTINUE
        WRITE (*, 10001)
     +  'OEnter operation (A-add, C-change, D-delete, L-list)? '
10001   FORMAT (A, \)
        READ (*, 10002) RESPON
10002   FORMAT (A)
C     A. If Add then ADD:
        IF (RESPON .EQ. 'A' .OR. RESPON .EQ 'a') THEN
            CALL ADDDB
C     B. If Change then CHANGE:
        ELSEIF (RESPON .EQ. 'C' .OR. RESPON .EQ. 'c') THEN
            CALL CNGDB
C     C. If Delete then DELETE
        ELSEIF (RESPON .EQ. 'D' .OR. RESPON .EQ. 'd') THEN
            CALL DELDB
C     D. If List then LIST
        ELSEIF (RESPON .EQ. 'L' .OR. RESPON .EQ. 'l') THEN
            CALL LSTDB
```

```
C        F. If invalid, display message and GOTO II
            ELSE
                WRITE (*,*) 'A/C/D/L are only valid responses'
                GOTO 100
            ENDIF
C III. Prompt/read any more food items?
            WRITE (*, 10001) ' Anything else? {Y/N}? '
            READ (*, 10002) RESPON
C IV. If more, GOTO II
            IF (RESPON .EQ. 'Y' .OR. RESPON .EQ. 'y') THEN
                GOTO 100
            ENDIF
C V. Shutdown and exit
            CALL SHTDWN

            END
**********************************************
C Initialize Data Base Files
            SUBROUTINE GETDB

C <> <> <> <> <> <> <> <> <> <> <> <> <> <> <> <>
C DIRectory COMMON
C                            DIR contains the record number into
C                            CAL_DAT
            INTEGER*2 DIR (997)
C                            LSTREC is the last record written in
C                            CAL_DAT
            INTEGER*2 LSTREC
C
            COMMON /DIRECT/ DIR, LSTREC

C                            THERE is .TRUE. if CAL_DIR.DAT exists
            LOGICAL THERE

C 1. INQUIRE status of DB file
            INQUIRE (FILE='CAL_DIR.DAT', EXIST=THERE)
            IF ( .NOT. THERE ) THEN
C       a. If none, OPEN CAL_DAT.DAT New and clear DIR
                OPEN (1, FILE='CAL_DAT.DAT', STATUS='NEW',
        +           ACCESS='DIRECT', FORM='UNFORMATTED', RECL=32)
                LSTREC = 0
                DO 100 I=1, 997
                    DIR(I) = 0
```

(continued)

```
100          CONTINUE
C     b. If one, OPEN CAL_DAT.DAT and OPEN and READ CAL_DIR.DAT and
C        CLOSE
         ELSE
            OPEN (1, FILE='CAL_DAT.DAT', STATUS='OLD',
     +          ACCESS='DIRECT', FORM='UNFORMATTED', RECL=32)
            OPEN (2, FILE='CAL_DIR.DAT', STATUS='OLD',
     +          ACCESS='SEQUENTIAL', FORM='UNFORMATTED')
            READ (2) (DIR(I), I=1, 997), LSTREC
            CLOSE (2)
         ENDIF

         END
*******************************************************
C Add entry to Database
         SUBROUTINE ADDDB

C                         ERRCOD is error code.  0=OK
         INTEGER ERRCOD
C                         FOOD is the food item to be processed
         CHARACTER*20 FOOD
C                         FODLEN is the length of FOOD
         INTEGER FODLEN
C                         HASHCD is the Hash coded index of FOOD
         INTEGER HASHCD
C                         UNAME is the calorie unit's name
         CHARACTER*10 UNAME
C                         UVALUE is the calories per unit
         INTEGER*2 UVALUE
C                         RESPON is the user response
         CHARACTER RESPON

100      CONTINUE
C 1. Prompt/read food-item
         WRITE (*, 10001) '0Enter name of food: '
10001    FORMAT (A, \)
         READ (*, 10002) FOOD
10002    FORMAT (A)
C 2. Validate, count and convert to UC
         CALL FODITM (FOOD, FODLEN)
C 3. Hash code FOOD
         CALL HASHIT (FOOD, FODLEN, HASHCD)
```

```
C 4. If present already, display message
        CALL GETREC ( FOOD, HASHCD, UNAME, UVALUE, ERRCOD )
        IF ( ERRCOD .EQ. 0 ) THEN
            WRITE (*,*) 'That is already in the database.'
C 5. Else, add the new record
        ELSE
C 6. Prompt/read units
            WRITE (*, 10001) ' Enter unit name: '
            READ (*, 10002) UNAME
            WRITE (*, 10001) ' Enter calories in one unit: '
            READ (*, 10003) UVALUE
10003       FORMAT ( I4 )
C 7. WRITE out food-item and units into DB
            CALL ADDREC ( HASHCD, FOOD, UNAME, UVALUE, ERRCOD )
C 8. If full, display message
            IF ( ERRCOD .NE. 0 ) THEN
                WRITE (*,*) 'There is no room left in the database'
            ENDIF
        ENDIF
C 9. Ask if more Adds ...
        WRITE (*, 10001) ' More Adds {Y/N}? '
        READ (*, 10002) RESPON
        IF (RESPON .EQ. 'Y' .OR. RESPON .EQ. 'y') THEN
            GOTO 100
        ENDIF

        END
*****************************************************
C Change entry in Database
        SUBROUTINE CNGDB

        WRITE (*,*) 'Change not implemented'

        END
*****************************************************
C Delete entry in Database
        SUBROUTINE DELDB

        WRITE (*,*) 'Delete not implemented'

        END
```

(continued)

```
***************************************************
C List entry in Database
        SUBROUTINE LSTDB

        WRITE (*,*) 'List not implemented'

        END
***************************************************
C Close down the files and such

        SUBROUTINE SHTDWN
C <> <> <> <> <> <> <> <> <> <> <> <> <> <> <> <>
C DIRectory COMMON
C                       DIR contains the record number into
C                       CAL_DAT
        INTEGER*2 DIR (997)
C                       LSTREC is the last record written in CAL_DAT
        INTEGER*2 LSTREC
C
        COMMON /DIRECT/ DIR, LSTREC

C                       THERE is .TRUE. if CAL_DIR.DAT exists
        LOGICAL THERE

C 1. Display DB statistics
        WRITE (*,*) LSTREC, ' Records in the DB.'
C 2. INQUIRE status of DIR file and OPEN appropriately
        INQUIRE (FILE='A:CAL_DIR.DAT', EXIST = THERE)
        IF ( .NOT. THERE ) THEN
            OPEN (2, FILE='A:CAL_DIR.DAT', STATUS='NEW' ,
     +          ACCESS='SEQUENTIAL', FORM='UNFORMATTED')
        ELSE
            OPEN (2, FILE='A:CAL_DIR.DAT', STATUS='OLD',
     +          ACCESS='SEQUENTIAL', FORM='UNFORMATTED')
        ENDIF
C 3. WRITE the DIRectory
        WRITE (2) (DIR(I), I=1, 997), LSTREC
C 4. CLOSE Dir and DB
        CLOSE (2)
        CLOSE (1)

        END
```

```
*****************************************************
        SUBROUTINE FODITM (FOOD, FODLEN)
C See CALORY
        END
*****************************************************
        SUBROUTINE HASHIT (FOOD, FODLEN, HASHCD)
C See CALORY
        END
*****************************************************
C ADDREC will add an entry into the Database

        SUBROUTINE ADDREC ( HASHCD, FOOD, UNAME, UVALUE, RESCOD )

C                       HASHCD {I} ranges from 1 ... 997
        INTEGER HASHCD
C                       RESCOD {O} is the result code. 0=Add
C                       succeeded
        INTEGER RESCOD
C                       FOOD {I} is the food item
        CHARACTER*20 FOOD
C                       UNAME {I} is the calorie unit's name
        CHARACTER*10 UNAME
C                       UVALUE {I} is the calories per unit
        INTEGER*2 UVALUE
C <> <> <> <> <> <> <> <> <> <> <> <> <> <> <> <>
C DIRectory COMMON
C                       DIR is contains the record number into
C                       CAL_DAT
        INTEGER*2 DIR (997)
C                       LSTREC is the last record written in
C                       CAL_DAT
        INTEGER*2 LSTREC
C
        COMMON /DIRECT/ DIR, LSTREC

C 1. Initialize loop to HASHCD
        I = HASHCD
C 2. Loop to examine DIR entry
100     CONTINUE
C    a. If entry not empty
        IF ( DIR(I) .NE. 0 ) THEN
```

(continued)

```
C            1. increment to next entry in DIR - MOD 997
                I = I + 1
                IF ( I .GT. 997 ) THEN
                    I = 1
                ENDIF
C                WRITE (*,*) 'ADD Collision - try', I
C            2. If looped back to HASHCD, set RESCOD to "full" and
C            return
                IF ( I .EQ. HASHCD ) THEN
                    RESCOD = -1
                    RETURN
                ENDIF
                GOTO 100
            ENDIF
C      b. If located an empty entry in DIR
C            1. Increment LSTREC and
            LSTREC = LSTREC + 1
C            2. write FODREC data into LSTREC and enter that into DIR
            WRITE (1, REC=LSTREC) FOOD, UNAME, UVALUE
C            WRITE (*,*) 'Entry made at', I, ' for record', LSTREC
            DIR ( I ) = LSTREC
C            3. set RESCOD to ''success''
            RESCOD = 0

            END
*******************************************************
            SUBROUTINE GETREC ( FOOD, HASHCD, UNAME, UVALUE, RESCOD )
C See CALORY
            END
```

8. Extend CALORY to "remember" the meal, so an entire day's input can be reviewed.

9. Extend Exercise 8 to plot calories input over time. Users could select the time period they want to be plotted.

10. Extend Exercise 8 to "remember" all the meals for everyone in the family, say 20 people.

11. Normally, people eat three times a day—breakfast, lunch, and supper. Extend Exercise 8 or 10 to include the time of the meal; then when the meal is entered, your program could tell if any meals were skipped.

12. Extend CALDBM in Exercise 7 to include the CNGDB function to change an existing entry in the database.

13. Extend CALDBM in Exercise 7 to include the LSTDB function to list the database.

14. Extend CALDBM in Exercise 7 to include the DELDB function to delete an entry in the database. This is really a difficult problem because the entry following the deleted one—if there is one—must be rehashed to see if it belongs where it is or if it is part of the hash chain. Likewise, the one after that must also be rehashed until there is a gap in the chain. To do this problem justice, you must find another textbook that discusses hashing; Knuth is an excellent choice.

APPENDIXES

APPENDIX A

TABLE OF GENERIC INTRINSIC FUNCTIONS

FORTRAN has many "intrinsic," or built-in, functions. They are built-in in the sense that it is not necessary to declare them before using them, and they may be used in various contexts. For instance, ABS (absolute value) can be used with either a REAL or an INTEGER input variable, and the FUNCTION is automatically "declared" to agree in type with the input used, so that you get an INTEGER output when you use an INTEGER as input. Thus, ABS is called a "generic" intrinsic because the proper ABS function is selected by the compiler for the context of its usage.

This Appendix includes the complete list of FORTRAN 77 generic intrinsic FUNCTIONs. Many implementations of FORTRAN 77 also include additional intrinsics. Be aware that in using "nonstandard" intrinsics, you are creating a program that is no longer portable. This is a disadvantage, since the primary reason for the standard is portability.

A few interesting inconsistencies in this table are worth pointing out:

- SIN, COS, and LOG are all defined for COMPLEX data, yet LOG10 and TAN are not.
- Truncation and rounding functions are scattered over several functions, all ending with "INT," and it will certainly take you a long time to remember them all—if you care to.

These and other inconsistencies can be very disconcerting, prone to errors, and difficult to use generally—so beware. In FORTRAN's defense, these intrinsics have grown up over several generations of FORTRANs, as the need arose. So that old programs would continue to compile, the older intrinsics were never removed as new ones were added, and the whole package was never designed as a unit.

It is difficult to decide how to arrange such a table, alphabetically or by function, so we have done it both ways. If you know the name of the function, you can go directly to the second table. If you know only what you need generally, you may use the first table as index to the second.

Table A.1   General functions

General function	Possible function name
Type Conversion:	AIMAG, AINT, ANINT, CHAR, CMPLX, DBLE, ICHAR, INT, NINT, REAL
Misc. Mathematics:	ABS, DIM, MOD, SIGN, SQRT
Sorting:	MAX, MIN
Character:	CHAR, ICHAR, INDEX, LEN, LGE, LGT, LLE, LLT
Complex:	ABS, AIMAG, CMPLX, CONJG, DPROD, REAL
Trigonometric:	ACOS, ASIN, ATAN, ATAN2, COS, COSH, SIN, SINH, TAN, TANH
Logarithms:	EXP, LOG, LOG10
Hyperbolic:	COSH, EXP, SINH, TANH

Table A.2   Generic functions

Name	Function definition	In	Out	Example
ABS	Absolute value	I R D C	I R D R	ABS(−5) returns 5 ABS(7.7) returns 7.7 If CPX = (3.0,4.0) then ABS(CPX) returns 5.0
ACOS	Arc Cosine—radians	R D	R D	ACOS(−1) returns π
AIMAG	Imaginary part of COMPLEX number	C	R	If CPX = (1.2, 3.4) then AIMAG(CPX) returns 3.4
AINT	Truncate REAL or DOUBLE	R D	R D	AINT(3.5) returns 3.0
ANINT	Round REAL or DOUBLE	R D	R D	ANINT(6.7) returns 7.0
ASIN	Arc Sine—radians	R D	R D	ASIN(1) returns $\pi/2$
ATAN	Arc Tangent—radians	R D	R D	ATAN(1.0) returns $\pi/4$
ATAN2	Arc Tangent—radians	R D	R D	ATAN2(−1.0, −1.0) returns $\pi/4$
CHAR	Convert INTEGER to	I	Ch	CHAR(65) returns 'A.' This is provided to circumvent type checking
CMPLX	Conversion to COMPLEX	I R D C	C C C C	CMPLX(1,2) returns the COMPLEX value (1, 2)
CONJG	COMPLEX conjugate	C	C	If CPX = (9.8, 7.6) then CONJG(CPX) returns (9.8, −7.6)
COS	Cosine—radians	R D C	R D C	COS($\pi/2$) returns 0
COSH	Hyperbolic Cosine	R D	R D	COSH(0.5) returns 1.1276. . .

(continued)

Table A.1   (*continued*)

Name	Function definition	In	Out	Example
DBLE	Conversion to DOUBLE	I R D C	D D D D	DBLE(2) returns the double precision value 2.000000000
DIM	Positive difference	I R D	I R D	DIM(x,y) returns $x-y$ if $x>y$ and zero otherwise
DPROD	DOUBLE product of 2 REAL inputs	R	D	DPROD(3.0, 4.0) returns 12.0 in double precision format
EXP	Raise "e" to a power	R D C	R D C	EXP(0.5) returns 1.6487. . .
ICHAR	Convert CHARACTER to	Ch	I	ICHAR('Z') returns 90. This is provided to circumvent type checking
INDEX	Locate substring	Ch	I	INDEX('abcdef', 'cd') returns 3
INT	Truncate and convert to INTEGER	R D C	I I I	INT(1.57) returns 1
LEN	Length of CHARACTER string	Ch	I	If the program contains CHARACTER*6 B then LEN(B) returns 6
LGE	Compare CHARACTER strings greater or equal	Ch	L	LGE('X', 'Y') returns .FALSE.
LGT	Compare CHARACTER strings greater than	Ch	L	LGT('MN', 'AB') returns .TRUE.
LLE	Compare CHARACTER strings less or equal	Ch	L	LLE('123 ', '123') returns .TRUE.
LLT	Compare CHARACTER strings less than	Ch	L	LLT('pppq', 'pppp') returns .FALSE.

Name	Function definition	In	Out	Example
LOG	Natural Logarithm	R D C	R D C	LOG(3.8) returns 1.335. . .
LOG10	Common Logarithm	R D	R D	LOG10(3.8) returns 0.57978. . .
MAX	Maximum of "n" arguments	I R D	I R D	MAX(1, 3, 2, 5, 0) returns 5
MIN	Minimum of "n" arguments	I R D	I R D	MIN(1, 3, 2, 5, 0) returns 0
MOD	Remainder after division	I R D	I R D	MOD(14, 3) returns 2
NINT	Round and convert to integer	R D C	I I I	NINT(1.57) returns 2
REAL	Conversion to REAL	I R D C	R R R R	REAL(5) returns 5.0 If CPX = (9.8, −7.6) then REAL(CPX) returns 9.8
SIGN	Transfer sign of first argument to the second	I R D	I R D	SIGN(−10, 1) returns −1
SIN	Sine—radians	R D C	R D C	SIN(π/2) returns 1.0
SINH	Hyperbolic Sine	R D	R D	SINH(1) returns 1.1752. . .
SQRT	Square root	R D C	R D C	SQRT(81) returns 9
TAN	Tangent—radians	R D	R D	TAN(0.78539) returns 0.999. . .
TANH	Hyperbolic Tangent	R D	R D	TANH(1.5) returns 0.90514. . .

A final note of explanation. This presentation has been greatly simplified in order to make it more usable. When you run across a table like this in another reference, you will notice that every generic also has "specific" names. For instance, the generic ABS is also known as DABS for a DOUBLE PRECISION argument and CABS for a COMPLEX argument. Although the programmer may reference the built-in functions by their specific names, this option is unnecessarily complicated; therefore this information was omitted from the table.

APPENDIX B

COMPLETE SUMMARY OF FORTRAN STATEMENTS

This appendix contains a summary of the complete FORTRAN 77 statements. Many of these statements were not discussed in the text, but are included here for reference only. This is not intended to be a "stand-alone" document. It is included to provide the student with an introduction to the complete FORTRAN 77 language. For the most part, the notation used in this section is the same as used in the text:

- Special characters and upper case words are to be written as shown.
- Lower case words indicate general entities for which specific information must be supplied by the programmer.
- Braces, "{" and "}", are used to indicate optional items.

243

- An ellipsis, ". . .", indicates that the preceding optional item may be repeated.
- Blanks are used to improve readability, but generally have no significance.

ASSIGN label TO assign_var_name

BACKSPACE unit

BACKSPACE (backspace_list)

BLOCK DATA {name}

CALL subprogram_name {(arg$_1$ {, arg$_2$}. . .)}

CHARACTER {*length} var_name$_1$ {, var_name$_2$}. . .

CHARACTER {*(*)} var_name$_1$ {, var_name$_2$}. . .

CLOSE (close_list)

COMMON {/{block_name$_1$}/} var_list$_1$ {{,}/{block_name$_2$}/var_list$_2$}. . .

COMPLEX var_name$_1$ {, var_name$_2$}. . .

CONTINUE

DATA var_list$_1$ / const_list$_1$ / {{,} var_list$_2$ / const_list$_2$ /}. . .

DIMENSION var_name$_1$ (dim_list$_1$) {, var_name$_2$(dim_list$_2$)}. . .

DO label {,} var_name = expr$_1$, expr$_2$ {, expr$_3$}

DOUBLE PRECISION var_name$_1$ {, var_name$_2$}. . .

ELSE

ELSEIF (log_expr) THEN

END

ENDFILE unit

ENDFILE (endfile_list)

ENDIF

ENTRY entry_name {(({dummy$_1$ {, dummy$_2$}. . . })}

EQUIVALENCE (equiv_pair$_1$) {, (equiv_pair$_2$)}. . .

EXTERNAL sub_name$_1$ {, sub_name$_2$}. . .

FORMAT (field$_1$ {, . . .} {, field$_n$})

func_name ({dummy$_1$ {, dummy$_2$}. . . }) = expr

{type} FUNCTION func_name {(({dummy$_1$ {, dummy$_2$}. . .})}

GOTO assign_var_name {{,} (label$_1$ {, label$_2$}. . .)}

GOTO label

GOTO (label$_1$ {, label$_2$}. . .) {,} expr

IF (log_expr) statement

IF (log_expr) THEN

IF(expr) label$_1$, label$_2$, label$_3$

IMPLICIT type (var_name$_1$ {, var_name$_2$}. . .) {, type (var_name$_1$ {, var_name$_2$}. . .)}. . .

INQUIRE (file_list)

INQUIRE (unit_list)

INTEGER var_name$_1$ {, var_name$_2$}. . .

INTRINSIC func_name$_1$ {, func_name$_2$}. . .

LOGICAL var_name$_1$ {, var_name$_2$}. . .

OPEN (open_list)

PARAMETER (param_name$_1$ = param_expr$_1$ {, param_name$_2$ = param_expr$_2$}. . .)

PAUSE {expr}

PRINT form {, io_list}

PRINT * {, io_list}

PROGRAM prgm_name

READ (read_list) {io_list}

READ form {, io_list}

READ * {, io_list}

REAL var_name$_1$ {, var_name$_2$}. . .

RETURN {expr}

REWIND unit

SAVE var_name$_1$ {, var_name$_2$}. . .

STOP {expr}

SUBROUTINE sub_name {({dummy$_1$ {, dummy$_2$} . . .})}

WRITE (write_list) {io_list}

WRITE form {, io_list}

WRITE * {, io_list}

APPENDIX C

COMPLETE WRITE AND READ FORMAT SPECIFICATIONS

This appendix is an expansion of the FORMAT specifications introduced in Chapter 6. As in that chapter, this presentation is divided into two sections: WRITE and READ. This appendix is a complete reference. The reader is directed back to Chapter 6 for the proper usage of these specifications.

C.1 WRITE

CARRIAGE CONTROLS

The first character to be *printed* is actually the second one on the line. Keep this in mind when designing your WRITE FORMAT. These carriage controls are summarized in the table that follows.

Plus	(+)	Overprint.	Start the line over again without advancing the carriage.
Blank	( )	Single space.	Start on the next line.
Zero	(0)	Double space.	Skip a line, then begin on following line.
One	(1)	Top of page.	Skip to perforation, then begin at the line following it.
Others		Same as a blank.	

Of course, you must consider the destination device when choosing carriage controls. The "top of page" doesn't make much sense on a CRT terminal—nor does "overprinting." These concepts are illustrated in the first example in "Put It Together" and the exercises in Chapter 6.

FORMAT FIELD SPECIFICATIONS

The general form of a "field" (with several exceptions) is:

$$r \; f \; w \; . \; d$$

where the following definitions apply:

- "r" is a positive integer, the "repeat" count. It is optional; if missing, it is assumed to be unity;
- "f" is the "field" code, described below;
- "w" is a positive integer, the field "width";
- "." is required punctuation;
- "d" is a non-negative integer which stands for number of "digits" to the right of the decimal;

In the examples below, the symbol ƀ means "blank space."

A FIELD $\{r\}A\{w\}$

The A field descriptor transfers characters to the print line. If the field width "w" is wider than the definition of the corresponding variable, the leading blanks are supplied. If "w" is smaller, only the leftmost "w" characters are transferred. If "w" is not present, the defined length of the corresponding variable or constant is used. Therefore, if "w" is not used, there is no way to make an error using the A field descriptor.

Field	Memory value	Output	Comments
A6	Ohms	ƀƀOhms	Leading blanks are added
A5	Volts	Volts	Exactly right
A4	Amperes	Ampe	Not enough room, the string is chopped off on the right

D FIELD   {r}Dw.d

The D field descriptor converts DOUBLE PRECISION variables in the same way that the E field descriptor converts REAL variables. The only difference is the optional *Ee* part of the E field isn't available on the D field. Notice that there is no counterpart to the F field for DOUBLE PRECISION variables.

E FIELD   {r}Ew.d{Ee}

The E field descriptor causes rounding, conversion, and transfer of REAL and COMPLEX variables to the output line in scientific, or exponential, notation. The total field width of the number is defined by "*w*" and the number of digits to the right of the decimal point is defined by "*d*." The exponent requires "*e*"+2 positions, and if "*e*" is not specified, it is assumed to be 2. Therefore, "*w*" must be greater than or equal to "*d*"+"*e*"+5. The converted number is right justified in the "*w*" positions, with leading blanks added if necessary.

Field	Memory value	Output	Comments
E9.1	123.4567123	ƀƀ0.1E+02	Notice how many significant digits were lost in the conversion
E11.4	123.4567123	ƀ0.1235E+02	Rounding takes place on the right-most digit
E12.2	−0.0000045	ƀƀƀ−0.45E−05	The decimal can move right too
E6.2	−0.0000045	******	Error—not enough room
E15.6E4	−2.0020000	−0.200200E+0001	The exponent is controlled by the E4 option

F FIELD   {r}Fw.d

The F field descriptor causes rounding, conversion, and transfer of REAL and COMPLEX variables to the output line in decimal notation. The total field width is defined by "w" and the number of digits to the right of the decimal point is defined by "d." Therefore "w" must be greater or equal to "d" + 3. The digits are right justified in the "w" positions, with leading blanks if necessary.

Field	Memory value	Output	Comments
F8.3	2.4567654	ƀ ƀ ƀ 2.457	Value is rounded, leading spaces
F7.2	− 987.1234567	− 987.12	Exactly enough room
F7.3	− 987.1234567	* * * * * * *	Error—not enough room
F8.2	− .2	ƀ ƀ ƀ − 0.20	Notice zero left of decimal

G FIELD   {r}Gw.d{Ee}

The G field descriptor combines features of the E and F field descriptors by determining, by the magnitude of the number, which descriptor is the "best." "Best" is determined by the following algorithm: If the number is less than 0.1 or larger than 10**"d," then an E format is used; otherwise an F format is used. The optional "Ee" part of this descriptor applies only if the E descriptor is chosen; see the E discussion to see how this works. Unlike the E and F descriptors, the converted number is not necessarily right justified in the "w" positions as illustrated below.

Field	Memory value	Output	Comments
G13.6	0.01234567	ƀ 0.123457E − 01	Number less than 0.1
G13.6	− 0.12345678	− 0.123457ƀ ƀ ƀ ƀ	Trailing blanks added
G13.6	123456.78901234	ƀ123457.ƀ ƀ ƀ ƀ ƀ ƀ	Leading and trailing blanks
G13.6	− 1234567.890123	− 0.123457E + 07	Number's magnitude larger than 10**6

I FIELD {r}Iw{.m}

The I field descriptor converts and transfers INTEGERs to the output line. The digits will be right justified in the "w" positions. If the number does not fill the "w" positions, leading blanks are transferred. The optional "m" field causes leading zeros (not blanks) in the rightmost "m" positions; "m" can be as large as "w", but if "m" is smaller, "w" − "m" are blank filled.

FIELD	Memory value	Output	
I3	456	456	
I2	234	**	Error—not enough room
I4	− 123	− 123	
I5	987	ƀƀ987	Two leading blanks are inserted
I4.2	6	ƀƀ06	A leading zero and two leading blanks are inserted
I3.3	− 65	− 65	The sign takes the place of the leading zero

L FIELD {r}Lw

The L field descriptor converts and transfers LOGICALs to the output line. There are only two possibilities: a T (true) or an F (false). Leading blanks are supplied if "w" is larger than 1.

FIELD	Memory value	Output
L4	.TRUE.	ƀƀƀT
L1	.FALSE.	F

P FIELD nP

The P field descriptor allows the programmer to alter the location of the decimal point in REAL, COMPLEX, and DOUBLE PRECISION variables. The P field is always used with either the D, E, F, or G fields. In effect, the internal number is scaled by 10**"n" before it is converted, where − 127 < = "n" < = 128. In an F field, the decimal is moved left or right. In an E or D field, the decimal point is moved and the exponent is changed by "n" also. Since the G field already scales its data, the effect of the P field is suspended in most cases. The nP scale factor stays in effect for the rest of the FORMAT statement or until a new nP descriptor is encountered.

Field	Memory value	Output	Comments
E12.3	-234.56	ƀƀ$-0.235E+03$	Without P field
1PE12.3	-234.56	ƀƀ$-2.346E+02$	P field adds one more digit to output without changing value
F6.1	987.21	ƀ987.2	Without P field
-2PF6.1	987.21	ƀƀƀ9.9	Decimal is moved and significance is lost

S FIELD   S

The S field descriptor operates like the SS field.

SP FIELD   SP

If a number is negative, the minus sign is always printed immediately to the left of the number—there is no programmer control over the minus sign. If the number is not negative, unless told otherwise, the plus sign is not printed. But the plus sign is under programmer control through the S, SP, and SS field descriptors. The SP descriptor causes leading plus signs to be placed to the left of the numbers. Plus signs will continue to be placed until either the S or SS field is encountered, or the end of the FOR-MAT is reached. The SP field is not matched to data.

Field	Memory value	Output	Comments
I6	87632	ƀ87632	Normal result
SP,I6	87632	$+87632$	Plus sign is added to output

SS FIELD   SS

The SS descriptor disables the SP descriptor action, which means that the plus sign will not be printed. The SS field is not matched to data. The SS is normally in effect at the beginning of the FORMAT.

T FIELD   Tn

The T field descriptor defines a position, "n", to put the next output character. No actual output is generated with this descriptor. Notice that the carriage control character is considered position 1, so the first printable character position is 2. This descriptor, in effect, allows the position pointer to move both right and left on the output line. All positions are absolute positions, rather than relative as in the case of the TL, TR, and X descriptors. Overprinting cannot be performed with this descriptor.

TL FIELD TL*n*

The TL field descriptor "tabs" the output position pointer "*n*" places to the left, without transferring any data to the line. This is, in effect, a relative tabulator. The TL works in the opposite direction of the TR descriptor. Overprinting cannot be performed with this descriptor.

TR FIELD TR*n*

The TR field descriptor "tabs" the output position pointer "*n*" places to the right, without transferring any data to the line. This is, in effect, a relative tabulator. The TR works in the opposite direction of the TL descriptor, and it works exactly like the X descriptor. Overprinting cannot be performed with this descriptor.

X FIELD *n*X

The X field descriptor tabs "*n*" positions to the right; in other words, it works just like the TR descriptor. There is no output for this descriptor, and overprinting cannot be performed with it.

Slash FIELD /

The slash terminates output of the current line and initiates a new one. It can also be used as carriage control, since a double slash causes one line to be skipped. If a slash is used, the comma separator is optional, but you will notice that commas make the FORMAT more "readable." The slash isn't required as the final character of a FORMAT—the right parenthesis will terminate the line—but it is permitted.

C.2 READ

The major difference in the operation of the READ/FORMAT over the WRITE/FORMAT is that data is being transferred *from* character form *to* internal binary form. We can still talk about "*w*," the width of the character field, but "*d*," the number of characters right of the decimal, has a new meaning which will be described shortly. There are other differences too, so stay alert. Normally, when looking at a FORMAT statement, it is difficult to tell if it belongs to a WRITE or a READ statement.

A FIELD {r}A{w}

The A field descriptor transfers "*w*" characters into a CHARACTER variable. Characters are transferred from the *left*—that is, the first character to be transferred is the leftmost. Then, "*w*" is used to determine how many characters to move to memory. If "*w*" is omitted, the declaration of the variable is used to find "*w*."

Field	Input	Memory value	Comments
A5	Windows	Windo	Input chopped off
A7	Windows	Windows	All characters transferred

BN FIELD BN

The BN descriptor is used immediately before a numeric descriptor (D, E, F, G, or I) to cause all embedded and trailing blanks in the number to be ignored. The BN descriptor stays in effect until either a BZ descriptor or the end of the FORMAT is encountered. The BN field is in effect at the beginning of the FORMAT.

Field	Input	Memory value
BN, I4	ƀ51ƀ	51
BN, I4	ƀƀ51	51
BN, I4	51ƀƀ	51
BN, I4	5ƀƀ1	51

BZ FIELD BZ

The BZ descriptor is used immediately before a numeric descriptor (D, E, F, G, or I) to cause all embedded and trailing blanks in the number to be treated as zeros. The BZ descriptor stays in effect until the end of the FORMAT or until a BN descriptor is encountered.

Field	Input	Memory value	Comments
BZ, I4	ƀ51ƀ	510	Trailing blank treated as a zero
BZ, I4	ƀƀ51	51	
BZ, I4	51ƀƀ	5100	
BZ, I4	5ƀƀ1	5001	All blanks become zeros

D FIELD {r}Dw.d

The D field descriptor converts "*w*" characters to DOUBLE PRECISION and stores the resulting value in memory. Any decimal notation is acceptable but, if the decimal is missing, "*d*" digits to the right of the decimal are assumed.

Field	Input	Memory value	Comments
D6.2	12.4ƀ ƀ	12.4D + 00	Decimal is used
D6.2	ƀ 1234ƀ	123.4D + 00	Decimal is assumed from right of field
D10.3	1234E + 5ƀ ƀ ƀ	1.234D + 05	Decimal is assumed from the right of E
D10.0	1234D − 7ƀ ƀ ƀ	1234.0D − 7	No decimal assumed

E FIELD {r}Ew.d{Ee}

The E field descriptor converts "*w*" characters to REAL and stores the resulting value in memory. Any decimal notation is acceptable but, if the decimal is missing, "*d*" digits to the right of the decimal are assumed. The optional "*Ee*" part has no meaning in a READ operation, but is accepted as valid. Notice that there is no difference in the operation of the E, F, and G fields when used with a READ statement.

Field	Input	Memory value	Comments
E6.2	12.4ƀ ƀ	12.4E + 00	Decimal is used
E6.2	ƀ 1234ƀ	123.4E + 00	Trailing zeros are assumed
E10.3	1234E + 5ƀ ƀ ƀ	1.234E + 05	Decimal is assumed from the right of the E
E10.3	ƀ 12.34D − 7ƀ	12.34E − 07	D notation is assumed to be E notation
E10.3E4	ƀ − 12.3E2ƀ ƀ	− 12.3E + 02	E4 notation ignored

F FIELD {r}Fw.d

The F field descriptor converts "*w*" characters to REAL and stores the resulting value in memory. Any input decimal notation is acceptable but, if the decimal is missing, "*d*" digits to the right of the decimal are assumed.

Notice that there is no difference between the E, F, and G fields when used with a READ statement.

Field	Input	Memory value	Comments
F7.4	1234567	123.4567	Decimal is assumed
F7.4	12345ƀ ƀ	123.45	Trailing zeros are assumed
F7.4	ƀ − 12.3ƀ	− 12.3	Decimal is used
F7.4	−6.2E3ƀ	−6200.0	E notation accepted
F7.4	−6.23D5	−623000.0	D notation accepted

G FIELD   {r}Gw.d{Ee}

The G field descriptor operates exactly the way the E and F fields operate. It is included in FORTRAN to be compatible with WRITE/FORMAT statements.

I FIELD   {r}Iw{.m}

The I field descriptor converts "w" characters to INTEGER and stores the resulting value in memory. The character string must be in INTEGER form; it cannot contain a decimal point or an exponent field. The optional ".m" part is ignored if present; it is accepted by FORTRAN so the FORMAT can also be used in a WRITE statement.

Field	Input	Memory value	Comments
I4	− 12ƀ	− 12	Trailing blanks are ignored
I5	ƀ ƀ 97ƀ	97	The number can be anywhere in the "w" character field

L FIELD   {r}Lw

The L field descriptor converts "w" characters to LOGICAL. If the leftmost nonblank(s) in the "w" character field are T, t, .T or .t, then a value of .TRUE. is assigned to the corresponding memory variable. If the leftmost nonblank(s) are F, f, .F or .f, then a value of .FALSE. is assigned to the variable. Notice that the exact spelling of "true" or "false" is not required; the first letter determines the value.

Field	Input	Memory value	Comments
L5	b̸ b̸ tb̸ b̸	.TRUE.	Letter can be anywhere in the field
L7	.FLASE.	.FALSE.	Misspelling not detected

P FIELD   nP

The P field descriptor defines a scale factor to be applied to a D, E, F, or G field descriptor. The number in the field is converted as defined above for the descriptor and then multiplied by $10**(-n)$ before memory storage. There is one exception to the application of the P field: When the character string contains an explicit exponent, the scale factor is NOT applied.

Field	Input	Memory value	Comments
2PF6.3	12.3b̸ b̸	0.123	Decimal point moved left 2 places
−3PE6.3	b̸.123b̸	123.0	Decimal moved right 3 places
4PG8.3	−0.654E4	−0.654E4	Decimal not moved

T FIELD   Tn

The T field descriptor defines a position, "n," for the location of the next input character. No variable in the I/O list is associated with this field. This descriptor allows the programmer to skip around in the input line or to reprocess a field rather than to process it in strictly left-to-right order. Notice that, unlike output, the input line has actual data in column 1, the leftmost column.

TL FIELD   TLn

The TL field descriptor moves the position pointer left "n" columns. This works opposite to the TR descriptor described below. It is, in effect, a "relative tabulator," whereas the T descriptor is an "absolute tabulator." This descriptor allows the programmer to skip fields while moving backward through the input line.

TR FIELD   TRn

The TR field descriptor moves the position pointer right "n" columns. This works opposite to the TL descriptor described above, and works exactly like the X descriptor. It is, in effect, a "relative tabulator," whereas the T descriptor is an "absolute tabulator." This descriptor allows the programmer to skip fields while moving forward through the input line.

X FIELD   nX

The X field descriptor works exactly like the TR descriptor described above.

Slash FIELD   /

The slash terminates input of the current line and reads a new one. Its primary purpose is to process more than one line with a single READ statement. If a slash is used, the comma separator is optional, but commas do make the FORMAT more "readable." The slash is not required as the first character of a FORMAT; the left parenthesis will cause a new line to be read.

APPENDIX D

CASE USING
A GOTO

This appendix contains an explanation for an alternative method to create a CASE statement in FORTRAN. This method is limited because the programmer can only test for integers—it is not as general as the technique discussed in Chapter 3. It also has the drawback of using the "indexed" GOTO construction; on the other hand, that is also its advantage.

The "indexed" GOTO is much faster than the IF-ELSEIF construction, especially when there are many alternative steps in the CASE. The student may judge the utility of this CASE form for him- or herself.

Consider the following general form:

```
           GOTO (label_list) expr
    label₁  CONTINUE
           statement(s)
           GOTO label

    label₂  CONTINUE
           statement(s)
           GOTO label

              . . .

    labelₙ  CONTINUE
           statement(s)

    label   CONTINUE
```

where the following definitions apply:

- "label_list" is an ordered list of labels of the form $label_1$, $label_2$, . . . , $label_n$. Labels in this list are separated by commas;
- "$label_i$" is a label;
- "expr" is a variable or an expression. If necessary it will be converted to an integer;
- "statement(s)" is one or more complete FORTRAN statements.

And this is how the pseudo CASE will work:

1. The expression "expr" is evaluated and converted to an integer, if necessary.
2. Assuming the value of the expression is computed to be "i," then the ith label from the "label_list" is used to perform a GOTO.
3. The group of statement(s) starting at "$label_i$" is executed, ending with the "GOTO label." The "GOTO label" guarantees that one and only one case selection is made.

Notice that the "$label_i$s" in "label_list" don't have to be unique, which means that two (or more) labels in the list can be identical. For instance, suppose that $label_j$ and $label_k$ are the same. Then, when "expr" is either j or k, the same set of statements will be executed.

You can readily see that the FORTRAN version is a subset of Pascal's CASE, where the following restrictions apply:

1. Only integers can be used for the case selector.
2. If the "expr" is less than one or larger than the number of labels in the "label_list," then the statement following the GOTO is executed. This means that the OTHERWISE—if you are familiar with that Pascal variation—is the first statement following the GOTO. Some versions of FORTRAN will warn you of this condition.

APPENDIX E

INPUT AND OUTPUT
OPERATIONS

This appendix is the companion to Chapter 10 in that it provides the details of the input and output operations described and illustrated in that chapter. It is organized into two major divisions: (1) sequential input and output operations and (2) direct input and output operations. Within each division, the operations are ordered alphabetically for reference purposes.

E.1 SEQUENTIAL FILE INPUT AND OUTPUT OPERATIONS

In the explanations of the statements you will notice several inconsistencies well worth flagging:

1. Some statements use parentheses and some don't.
2. Some arguments require a keyword, some don't.
3. Some arguments are INTEGERs and some are CHARACTER strings. CHARACTER string arguments must be enclosed in "tic" marks.

In the detailed descriptions of the operations that follow, definitions such as "unit" have been repeated whenever necessary. This saves you, the user/programmer, time looking for previously defined fields, even though it makes the section a bit longer. Again, to make them easier to locate, operations have been arranged alphabetically. Finally, arguments *must* be ordered as indicated below.

BACKSPACE

Consider the following model: As a file is accessed, the system maintains a "pointer" to remember where it is in the file. The purpose of the BACKSPACE statement is to backup the file pointer depending on the current position of the pointer.

- If the file pointer is at the beginning of the nth record, it is moved back to the beginning of the (n − 1)th record.
- If the file pointer is already at the beginning of the file, that is, at record 1, nothing happens.
- If the file pointer is past the end-of-file record, the pointer is moved to the beginning of the end-of-file record.
- If the file pointer is in the middle of a record, it is moved to the beginning of that same record.

The form of this statement is:

BACKSPACE unit

where the following definition applies:

- "unit" is a positive one- or two-digit integer expressed as a constant or stored in an INTEGER variable. "Unit" is a shorthand for the file name. The correspondence between "unit" and file name is made in the OPEN statement.

CLOSE

The CLOSE statement dissociates the file name/unit connection. It is fairly difficult to get an error on this statement.

The form of this statement is:

```
CLOSE (unit {,STATUS = stat} {,IOSTAT = check})
```

where the following definitions apply:

- "unit" is a positive one- or two-digit integer expressed as a constant or stored in an INTEGER variable. "Unit" is a shorthand for the file name. The correspondence between "unit" and file name is made in the OPEN statement.
- "stat" is either the string 'KEEP' or 'DELETE'. This tells the system what to do with your file, now that you have signaled you are done with it. If this argument does not appear, the default is 'DELETE' if the file was OPENed as a scratch file; otherwise the default is 'KEEP'.
- "check" is an INTEGER variable used to store the results of the CLOSE operation. Zero means that no error occurred. Other values depend on your operating system.

ENDFILE

The ENDFILE statement writes a special record on the file called the end-of-file record. When this is done, no more writes can take place until the file pointer is repositioned using the CLOSE, BACKSPACE, or REWIND statement. This record is used to exit a READ via the EOF argument.

The form of this statement is:

```
ENDFILE unit
```

where the following definition applies:

- "unit" is a positive one- or two-digit integer expressed as a constant or stored in an INTEGER variable. "Unit" is a shorthand for the file name. The correspondence between "unit" and file name is made in the OPEN statement.

INQUIRE

The INQUIRE statement determines all attributes of a file. It may be invoked any time: before OPEN, after OPEN, after READ/WRITE operations, and after CLOSE, but many of the parameters are valid only if the file is OPEN. Information about the file is provided according to the parameters that are supplied. There are two modes of inquiry, either by UNIT or by FILE. When INQUIREing by UNIT, the OPEN operation must have been performed, but INQUIREing by FILE can be done anytime.

The form of this statement is:

```
INQUIRE ({UNIT=unit} {FILE=fname} {,ERR=error_label}
{,EXIST=log_exist} {,NAMED=log_named} {,IOSTAT=check}
{,OPENED=log_opened} {,NUMBER=unit_numb} {,NAME=file_name}
{,ACCESS=acc_type} {,SEQUENTIAL=char_seq} {,DIRECT=char_direct}
{,FORM=form} {,FORMATTED=char_form} {,UNFORMATTED=char_unform}
{,RECL=rec_length} {,NEXTREC=next_numb} {,BLANK=bla}
```

where the following definitions apply (either "unit" or "fname" must be present—but not both—and all other parameters are optional):

- "unit" is a positive one- or two-digit integer expressed as a constant or stored in an INTEGER variable. If "unit" is specified, "fname" is not allowed.
- "fname" is the name of the file stored in as a CHARACTER variable or constant. You must follow your system rules for naming files. If "fname" is specified, "unit" is not allowed.
- "error_label" is a statement label number. If this parameter is specified and an error occurs during an INQUIRE operation, a GOTO "error_label" will take place.
- "log_exist" is a LOGICAL variable which is set to .TRUE. if the "fname" or the "unit" exists—depending on which is specified— and is set .FALSE. otherwise.

- "log_named" is a LOGICAL variable that is set to .TRUE. if the file was OPENed by using a file name—that is, if it is not a scratch file—and is .FALSE. otherwise. Notice that because this parameter applies only to OPENed files, it is used with the "unit" parameter only.
- "check" is an INTEGER variable used to store the results of the INQUIRE operation. Zero means that no error occurred. Other values depend on your operating system. If you don't supply this parameter and if an error occurs, your program will abort and the system will display an error message.
- "log_opened" is a LOGICAL variable that is set to .TRUE. if the "fname" has already been opened, and to .FALSE. otherwise. Notice that if the "unit" parameter is supplied the file *must* be OPEN already.
- "unit_numb" is an INTEGER variable in which the "unit" is stored if the file is already OPEN. "Unit_numb" is undefined if the file is not OPEN.
- "file_name" is a CHARACTER variable in which the "file" name is stored, which implies that the file is already OPEN.
- "acc_type" is a CHARACTER variable that is set to one of two strings: 'SEQUENTIAL' or 'DIRECT' if OPEN, and undefined otherwise.
- "char_seq" is a CHARACTER variable that is set to one of three strings: 'YES,' 'NO,' or 'UNKNOWN.' IF the file is OPEN and sequential, the "char_seq" variable is set to 'YES'; if the file is OPEN and not sequential, it is set to 'NO.' If the file is not OPEN, the "char_seq" is set to 'UNKNOWN.'
- "char_direct" is a CHARACTER variable that is set to one of three strings: 'YES,' 'NO,' or 'UNKNOWN.' If the file is OPEN and direct, the "char_direct" variable is set to 'YES'; if the file is OPEN and not direct, it is set to 'NO.' If the file is not OPEN, the "char_direct" is set to 'UNKNOWN.'
- "form" is a CHARACTER variable that is set to one of two strings: 'FORMATTED' or 'UNFORMATTED,' depending on the file's record type. The file must be OPEN to use this parameter successfully.
- "char_form" is a CHARACTER variable that is set to one of three strings: 'YES,' 'NO,' or 'UNKNOWN.' If the file is OPEN and formatted, the "char_form" variable is set to 'YES'; if the file is OPEN and not formatted, it is set to 'NO.' If the file is not OPEN, the "char_form" is set to 'UNKNOWN.'

- "char_unform" is a CHARACTER variable that is set to one of three strings: 'YES,' 'NO,' or 'UNKNOWN.' If the file is OPEN and unformatted, the "char_unform" variable is set to 'YES'; if the file is OPEN and not unformatted, it is set to 'NO.' If the file is not OPEN, the "char_unform" is set to 'UNKNOWN.'
- "rec_length" is an INTEGER variable that is set to the length of the direct file record length (in bytes, of course). The file must be OPEN to use this parameter correctly.
- "next_numb" is an INTEGER variable that is set to the record number of the next record to be accessed in the file; the first record of a file is always numbered 1. To use this parameter correctly, the file must be OPEN.
- "bla" is a CHARACTER variable which is set to one of two strings: 'NULL' or 'ZERO.' If the BN FORMAT descriptor is in effect, "bla" is set to 'NULL.' If the BZ FORMAT descriptor is in effect, "bla" is set to 'ZERO.' To use this parameter correctly, the file must be OPEN and FORMATTED. The BZ and BN FORMAT descriptors are defined in Appendix C.

OPEN

The OPEN statement performs several operations, as indicated in the arguments below:

- Creates an association between the file name and the unit so that only unit is needed in the other file operations.
- If the file is 'OLD,' confirms that the file exists and that the file characteristics—SEQUENTIAL/DIRECT and FORMATTED/UNFORMATTED—have not changed.
- If the file is 'NEW,' confirms that the file does not exist, then creates the file with the specified characteristics.
- Reports the results of the OPEN operation to the program.

Only when the OPEN operation is correctly performed will the program be allowed to do any other operations on the file. Not even CLOSE will work correctly if the file is OPENed incorrectly.

The form of the OPEN statement for sequential files is:

```
OPEN (unit {,FILE=fname} {,STATUS=status}
          {,FORM=format} {,IOSTAT=check} {,ERR=err_label})
```

where the following definitions apply:

- "unit" is a positive one- or two-digit integer expressed as a constant or stored in an INTEGER variable. "Unit" is a shorthand and is used in all the rest of the I/O statements instead of the file name.
- "fname" is the name of the file to be linked to "unit." You must follow your system rules for naming files; a dummy name will be created by OPEN if you don't specify it. Furthermore, if this argument is not present, the file is termed a "scratch" or temporary file and, unless you do something special, will be destroyed when CLOSE is executed.
- "status" is either the string 'NEW' or 'OLD' signifying to OPEN if it should expect "fname" to exist ('OLD') or not ('NEW'). This must be done right; trying to open a nonexistent file 'OLD' will result in an error. INQUIRE is useful to find out if the file exists or not.
- "format" can be either the string 'FORMATTED' or 'UNFORMATTED'; if not specified, it defaults to 'FORMATTED.' If you specify 'FORMATTED,' then you *must* supply a FORMAT statement for every READ/WRITE operation on the file.
- "check" is an INTEGER variable used to store the results of the OPEN operation. Zero means that no error occurred. Other values depend on your operating system. If you don't supply this parameter and if an error occurs, your program will abort and the system will display some sort of error message. If you do supply this parameter and an error occurs, your program maintains control—usually to display some meaningful, user friendly error message before stopping. The best practice is to test this variable in an IF statement immediately after the OPEN.
- "error_label" is an optional statement label number. If this parameter is specified and an error occurs during an OPEN operation, a GOTO "error_label" will take place. Then, in the code at "error_label," the general coding practice would be to test the "check" variable and to display an appropriate user friendly message. In some cases, error recovery may be possible. The specific error codes are dependent on the system you are using.

READ

The READ operation transfers data from a file on an external medium (disk, tape) to memory. The file must be OPEN before using READ.

One great option—even to terminal READing—is the use of the END parameter. It offers an elegant way of exiting an input loop. Most systems support a keyboard entry (CTRL-Z) to simulate an end-of-file condition. So you can instruct the user to "Input data or CTRL-Z." Your input code then loops unconditionally until the END—the user signals the end of input. This is an improvement over using a particular artificial input value such as 999999 to indicate the end.

The form of the READ statement for sequential files is:

```
READ (unit {,format} {,IOSTAT=check} {,END=eof_label}
      {,ERR=error_label}) io_list
```

where the following definitions apply:

- "unit" is a positive one- or two-digit integer expressed as a constant or stored in an INTEGER variable. "Unit" is a shorthand for the file name. The correspondence between "unit" and file name is made in the OPEN statement.

- "format" is the label of a FORMAT statement. Notice that this field is optional, and is required only for 'FORMATTED' files.

- "check" is an INTEGER variable used to store the results of the READ operation. Zero means that no error occurred. Other values depend on your operating system. To make use of this feature, you must also specify the "error_label" option; look at that description for a further explanation.

- "eof_label" is an optional statement label number. If this is specified and the end-of-file record is read (see ENDFILE), a GOTO "eof_label" will take place. If this parameter is not specified and an end-of-file record is read, a system error will occur and your program will be aborted. Most "sequential" file programs should depend on the existence of an end-of-file to control their operation.

- "error_label" is an optional statement label number. If this parameter is specified and an error occurs during a READ operation, a GOTO "error_label" will take place. Then, in the code at "error_label," the general coding practice would be to test the "check" variable and to display an appropriate user-friendly message.

- "io_list" has been defined in Chapter 6.

REWIND

The operation of this statement is trivial: Wherever the record pointer is, it will relocate to the beginning of the file. If the storage device is a tape rather than a disk, the tape will be physically repositioned.

The form of this statement is:

REWIND unit

where the following definition applies:

■ "unit" is a positive one- or two-digit integer expressed as a constant or stored in an INTEGER variable. "Unit" is a shorthand for the file name. The correspondence between "unit" and file name is made in the OPEN statement.

WRITE

The WRITE operation transfers data from memory to a file on an external medium (disk, tape). The file must be OPEN before using WRITE.

The form of this statement is:

WRITE (unit {,format} {,IOSTAT=check}

{,ERR=error_label})   io_list

where the following definitions apply:

■ "unit" is a positive one- or two-digit integer expressed as a constant or stored in an INTEGER variable. "Unit" is a shorthand for the file name. The correspondence between "unit" and file name is made in the OPEN statement.
■ "format" is the label of a FORMAT statement. Notice that this field is optional and required only for 'FORMATTED' files.

- "check" is an INTEGER variable used to store the results of the WRITE operation. Zero means that no error occurred. Other values depend on your operating system. To make use of this feature, you must also specify the "error_label" option.
- "error_label" is an optional statement label number. If this parameter is specified and an error occurs during a WRITE operation, a GOTO "error_label" will take place. Then, in the code at "error_label," the general coding practice would be to test the "check" variable and to display an appropriate user friendly message. In some cases, error recovery may be possible. The specific error codes are dependent on the system you are using.
- "io_list" has been defined in Chapter 6.

E.2 DIRECT-FILE INPUT AND OUTPUT OPERATIONS

The operations supported by FORTRAN specifically for direct files are much simpler, but because direct files can also be accessed sequentially, both methods for READ and WRITE have been incorporated into the parameter list.

In the explanations of the statements you will notice several inconsistencies well worth flagging:

1. Some statements use parentheses; some don't.
2. Some arguments require a keyword; some don't.
3. Some arguments are INTEGERs and some are CHARACTER strings. CHARACTER string arguments must be enclosed in "tic" marks.

Finally, the arguments *must* be ordered as indicated. The possible sequential file/record operations will now be presented alphabetically.

CLOSE

The CLOSE statement dissociates the file name/unit connection. It will also destroy the file if so specified. It is rather difficult to get an error on this statement. This is identical to sequential CLOSE.

The form of this statement is:

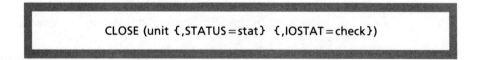

CLOSE (unit {,STATUS=stat} {,IOSTAT=check})

where the following definitions apply:

- "unit" is a positive one- or two-digit integer expressed as a constant or stored in an INTEGER variable. "Unit" is a shorthand for the file name. The correspondence between "unit" and file name is made in the OPEN statement.
- "stat" is either the string 'KEEP' or 'DELETE.' This tells the system what to do with your file, now that you have signaled you are done with it. If this argument does not appear, the default is 'DELETE' if the file was OPENed as a scratch file and 'KEEP' otherwise.
- "check" is an INTEGER variable used to store the results of the CLOSE operation. Zero means that no error occurred. Other values depend on your operating system.

INQUIRE

INQUIRE is the same for direct as for sequential files. Please refer to the discussion in Section E.1.

OPEN

The OPEN statement performs several operations, as indicated in the arguments above.

- Create an association between the file name and the unit so that only unit is needed in the other file operations.
- If the file is OLD, confirm that the file exists and that its characteristics—SEQUENTIAL/DIRECT, FORMATTED/ UNFORMATTED, and record length—have not changed.
- If the file is NEW, confirm that the file does not exist, then creates the file with the specified characteristics.
- Optionally, report the results of the OPEN operation to the program.

Only when the OPEN statement is correctly performed will the program be allowed to do *any* other operations in the file. Not even CLOSE will work correctly if the file is OPENed incorrectly.

The form of the OPEN statement for direct files is:

```
OPEN (unit {,FILE = fname} {,STATUS = status}

    ,ACCESS = 'DIRECT' {,FORM = format}

    {,IOSTAT = check} {,ERR = error_label} ,RECL = rec_length)
```

where the following definitions apply:

- "unit" is a positive one- or two-digit integer expressed as a constant or stored in an INTEGER variable. "Unit" is a shorthand used in all the other I/O statements instead of the file name.
- "fname" is the name of the file to be linked to "unit." You must follow your system rules for naming files: A dummy name will be created by OPEN if you don't specify one. Furthermore, if this argument is not present, the file is termed a "scratch" or temporary file and, unless you do something special, it will be destroyed when CLOSE is executed.
- "status" is either the string 'NEW' or 'OLD' signifying to OPEN whether it should expect "fname" to exist ('OLD') or not ('NEW'). This must be correct, since trying to open a nonexistent file 'OLD' will result in an error. INQUIRE is useful to find out if the file exists or not.
- "ACCESS = 'DIRECT'" is required to indicate that you are working with a direct file.
- "format" can either be the string 'FORMATTED' or the string 'UNFORMATTED,' and if it is not specified, defaults to 'UNFORMATTED.' If you specify 'FORMATTED,' you *must* supply a FORMAT statement for every READ/WRITE operation on the file.
- "check" is an INTEGER variable used to store the results of the OPEN operation. Zero means that no error occurred. Other values depend on your operating system. If you don't supply this parameter and if an error occurs, your program will abort and the

system will display an error message. If you do supply this parameter and an error occurs, your program maintains control— usually to display some meaningful, user friendly error message before stopping. The best practice is to test this variable in an IF statement immediately after the OPEN.

■ "error_label" is an optional statement label number. If this parameter is specified and an error occurs during an OPEN operation, a GOTO "error_label" will take place. Then, in the code at "error_label," the general coding practice would be to test the "check" variable and to display an appropriate user friendly message. In some cases, error recovery may be possible. The specific error codes are dependent on the system you are using.

■ "rec_length" is an integer value indicating the length of the record in bytes. This is required for direct files. The length of a record depends on the number and type of each of its variables, and you must make this computation. The algorithm is easy when you remember these three rules:

1. Each CHARACTER variable takes one byte. Of course a CHARACTER*10 variable takes ten bytes and arrays of CHARACTER variables must likewise be accounted for.
2. Each DOUBLE PRECISION or COMPLEX variable occupies eight bytes. If you have an array of either of these types, you must multiply the number of elements in the array by eight.
3. All other variables—REAL, INTEGER, and LOGICAL—take four bytes. Again, you must account for arrays by multiplying the number of elements in the array by four.

You aren't limited to these sizes; FORTRAN offers additional methods of saving space through another feature. LOGICAL and INTEGER variables can be declared to be smaller than four bytes using any of the following declarations:

1. LOGICAL*1 variables will occupy only one byte.
2. LOGICAL*2 variables will occupy two bytes.
3. INTEGER*2 variables will occupy two bytes. Of course, the maximum value that can be stored in these variables will necessarily be smaller than the standard INTEGER declaration you have been using. INTEGER*2 restricts you to $\pm 2^{15}$, that is, $\pm 32,767$.

INTEGER and INTEGER*4 are equivalent. Likewise, REAL and REAL*4 are the same; note that there is no REAL*2 because of the way the hardware represents real numbers. REAL*8 and DOUBLE PRECISION are the same. Now perhaps you can see an additional rationale for a declaration like CHARACTER*9—it falls right into the pattern.

READ

When using the "direct" READ operation to transfer a particular record to memory—say, record 497—you are implying that you already have a very good reason for picking 497 rather than 216. You know, at least in part, what is contained in that record. This is usually done by maintaining a second data structure that tells you what is contained in both record 216 and 497. This secondary structure is the "key" to locating data in the primary data file. To make this concept efficient, the secondary structure should be readily available in memory at all times; the primary structure is too large to fit in memory. So, your code first accesses the secondary structure with the "key" data to find the record number of the primary data.

In terms of our "address book" example in Chapter 10, the secondary structure may contain only the surnames of all the subscribers and—for each—the record number needed to access the complete data record in the primary data file for a subscriber. The primary data may consist of your friend's full name, address and phone number. Now, to make access fast, the surname structure may be ordered alphabetically so that a simple binary search may be performed. The order of the primary data then becomes immaterial.

The form of the READ statement for direct files is:

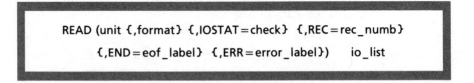

READ (unit {,format} {,IOSTAT=check} {,REC=rec_numb}
{,END=eof_label} {,ERR=error_label})   io_list

where the following definitions apply:

- "unit" is a positive one- or two-digit integer expressed as a constant or stored in an INTEGER variable. "Unit" is a shorthand for the file name. The correspondence between "unit" and file name is made in the OPEN statement.

- "format" is the label of a FORMAT statement. Notice that this field is optional and only required for "FORMATTED" files.
- "check" is an INTEGER variable used to store the results of the READ operation. Zero means that no error occurred. Other values depend on your operating system. To make use of this feature, you must also specify the "error_label" option; look at that description for a further explanation.
- "rec_numb" is the number of the record, starting at 1, that you are reading in the file. Of course you have noticed this is an optional parameter and think the book contains a typo: not true! When you specify REC=, that record is transferred to memory. If you *don't* specify REC=, a *sequential* READ takes place—that is, the next record from the file is transferred to memory.
- "eof_label" is an optional statement label number and makes sense to use only when READing the direct file sequentially. If this parameter is specified and the end-of-file record is read—see ENDFILE—a GOTO "eof_label" will take place. If this parameter is not specified and an end-of-file record is sequentially read, a system error will occur and your program will be aborted. If the REC= parameter is specified, there is no way to access the end-of-file record since it isn't numbered.
- "error_label" is an optional statement label number. If this parameter is specified and an error occurs during a READ operation, a GOTO "error_label" will take place. Then, in the code at "error_label," the general coding practice would be to test the "check" variable, and display an appropriate user friendly message. In some cases, error recovery may be possible. The specific error codes are dependent on the system you are using.
- "io_list" has been defined in Chapter 6.

WRITE

The "direct" WRITE operation transfers data from memory to a particular record or, as in "sequential" WRITE, to the "next" record. As in the case of READ, the underlying assumption is that you know what you are doing and have a secondary data structure to record "key" information about the record being written.

The form of this statement is:

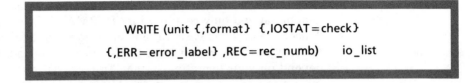

WRITE (unit {,format} {,IOSTAT=check}
{,ERR=error_label} ,REC=rec_numb) io_list

where the following definitions apply:

- "unit" is a positive one- or two-digit integer expressed as a constant or stored in an INTEGER variable. "Unit" is a shorthand for the file name. The correspondence between "unit" and file name is made in the OPEN statement.
- "format" is the label of a FORMAT statement. Notice that this field is optional; only "FORMATTED" files require it.
- "check" is an INTEGER variable used to store the results of the WRITE operation. Zero means that no error occurred. Other values depend on your operating system.
- "rec_numb" is the INTEGER number of the record—starting at 1— which you are writing out to the file. When you specify REC= rec_numb, that record is transferred from memory. If you *don't* specify REC=, a *sequential* WRITE takes place—that is, the next record is transferred to the file.
- "io_list" has been defined in Chapter 6.

BIBLIOGRAPHY

Borse, G. J., *FORTRAN 77 for Engineers*. Boston, Mass.: PWS Engineering, 1985.

Britton, Jack R., and L. Clifton Snively, *Algebra for College Students*. Rinehart & Company, Inc., 1954.

Burington, Richard Stevens, *Handbook of Mathematical Tables and Formulas*, 3rd ed. Handbook Publishers, Inc., 1957.

Forsythe, George E., *et al.*, *Computer Methods for Mathematical Computations*. Englewood Cliffs, N.J.: Prentice-Hall, 1977.

Johnson, Walter C., *Transmission Lines and Networks*. New York, N.Y.: McGraw-Hill, 1950.

Maron, Melvin J., *Numerical Analysis: A Practical Approach*. New York, N.Y.: Macmillan, 1982.

Microsoft FORTRAN Compiler for MS-DOS. Hewlett-Packard, 1985.

Radiotron Designer's Handbook, 4th ed. Radio Corporation of America, 1953.

Rule, Wilfred P., *FORTRAN 77: A Practical Approach*. Boston, Mass.: PWS Publishers, 1983.

Sears, Francis Weston, and Mark W. Zemansky, *University Physics*, 2nd ed. Reading, Mass.: Addison-Wesley, 1957.

Shames, James H., *Engineering Mechanics—Statics*. Englewood Cliffs, N.J.: Prentice-Hall, 1960.

Streeter, Victor L., *Fluid Mechanics*. New York, N.Y.: McGraw-Hill, 1958.

Van Valkenburg, M. E., *Network Analysis*. Englewood Cliffs, N.J.: Prentice-Hall, 1959.

Van Wylen, Gordon J., *Thermodynamics*. New York: N.Y.: John Wiley & Sons, 1962.

VAX-11 FORTRAN Language Reference Manual. Bedford, Mass. Digital Equipment Corporation, 1982.

INDEX

A 6
B 7
C 8
D 9
E 0
F 1
G 2
H 3
I 4
J 5